POWER TO CHANGE
THE WORLD

POWER TO CHANGE THE WORLD

Alternative Energy &

the Rise of the Solar City

by

S. L. Klein

BookSurge Publishing

Copyright © 2008 by S. L. Klein

All rights reserved. No part of this publication may be reproduced, distributed, or transmitted in any form or by any means without the prior written permission of the author with these exceptions: brief quotations embedded in critical reviews; certain noncommercial uses permitted by copyright law; photos, government graphs, and illustrations in the book that are in the public domain. For permission, contact the author through Booksurge.com.

BookSurge Publishing
7290 B Investment Drive
Charleston, South Carolina 29418

First BookSurge edition 2008

Printed in United States of America

Library of Congress Control Number: 2007909460

Library of Congress Cataloguing-in-Publication Data

S. L. Klein
Power to Change the World:
Alternative Energy & the Rise of the Solar City
Includes bibliographical references

International Standard Book Number (ISBN): 978-1-419-6798-7-2

Book cover designed by S.L. Klein from a photo by Derek Jensen

This book uses recycled paper.

To the State of California, Governor Arnold Schwarzenegger, his dedicated Staff, the California Legislature, and the companies here and around the world that are working to replace hydrocarbons with Alternative Energy

To
Robert Klein

And in memory of
Nikola Tesla

TABLE OF CONTENTS

PREFACE XI
ACKNOWLEDGEMENTS XIII

PART ONE
THE AGE OF HYDROCARBONS

1. THE AMAZING TWENTIETH CENTURY 3
2. PILLARS OF SOCIETY 25
3. A PLACE AT THE TABLE 30
4. DEPENDENCY AND PARADOX 33
5. OIL AND THE THIRD WAVE OF TOTALITARIANISM 34
6. THE ENVIRONMENT, THE MAD HATTER, AND HIS DIRTY DISHES 59
7. THE CLIMATE THRESHOLD 80
8. POPULATION GROWTH BLOW OUT 86
9. THE HIDDEN COSTS OF OIL 89
10. BACK TO THE FUTURE 104

PART TWO
AMERICA AT THE CROSSROADS

11. ALTERNATIVE TECHNOLOGIES 109
12. BUILDING BLOCKS AND STEPPING STONES IN TRANSPORTATION 111
13. BIOFUELS AND COAL: CATALYSTS OF TRANSITION? 132
14. BROADENING THE STATIONARY POWER BASE 138
15. A PARALLEL PATH TO THE FUTURE 156
16. THE HYDROGEN INTERNAL COMBUSTION ENGINE 158
17. THE PEM 172
18. HY-WIRE ACT 182
19. STATIONARY POWER FUEL CELLS 185
20. HYDROGEN PRODUCTION, DISTRIBUTION, AND STORAGE 190
21. TECHNOLOGICAL READINESS ASSESSMENT 205
22. THE LACK OF VISIBLE PROGRESS 206
23. WHAT HOLDS US BACK 209

PART THREE
THE AGE OF TRANSITION

24. CALIFORNIA BELLWETHER 229
25. THE FUTURE IN OUR HANDS 235
26. THE NEAR-TERM CAMPAIGN 252
27. THE ALTERNATIVE ENERGY ZONE 274
28. THE BOTTOM LINE 309
29. A CALL TO ACTION 313

MAJOR REFERENCES 315
END NOTES 317

Preface

In the 1970s, U.S. domestic oil production began a decline that has made energy independence a memory. Our dependency on others for so vital a commodity as oil results from our staggering consumption. To run modern American society, we require a quarter of the world's oil, about 21 million barrels a day. If this amount of oil were stored in one gallon cans, the cans set side by side would encircle the Earth at the equator approximately six times.[1] The need for so much oil at a time when our domestic oil production can no longer meet our needs has exposed us to increasing economic and social threat. The dependence of Europe, Japan, and Australia on oil imports is also dangerously high, all of which has placed Western democracy in jeopardy. What is at stake is suggested, if imperfectly understood, by Iran's theocratic "supreme leader" Ali Khamenei in a little noted 2002 address: "If they [the Western countries] do not receive oil," he said, "their factories will come to a halt. This will shake the world."[2]

Adding to the gathering political, social, and economic threat, our dependency on hydrocarbons has exposed us to an even darker prospect of severe ecological degradation. Simply put, the chemical alteration and combustion of hydrocarbons is poisoning the air we breathe, the water we drink, and the food we eat. The impact on the environment would be bad enough if we and the rest of the world simply maintained our current level of consumption, but we can't, given our present pace of expansion. The pressure of population growth and the clamor for the good life around the world have heightened the demand for oil and other hydrocarbons, which is accelerating the pace of degradation.

As a consequence of these multiple threats, a perfect storm is gathering over modern civilization, a converging political, social, economic, and ecological hydrocarbon catastrophe that must be mitigated before it is realized.[3]

When I began researching the material for this book, I worried about making the case for changing our energy consumption habits. Despite compelling evidence that we are endangering ourselves through our hydrocarbon consumption, little momentum existed for change. The average price of gasoline across the country was under $2.00. Except for an occasional blackout, just about everyone took our electrical grid for granted. Most of the broadcast media paid little attention to our declining domestic oil production. Relatively few seemed to care about our increasing dependency on foreign oil imports. Environmental degradation was largely ignored, and global warming was disputed.

Today, thankfully, perceptions are changing. Organizations, national labs, companies, the media, and individuals across our country and around the world are mobilizing in increasing numbers to fight the threats to our global energy infrastructure and environment. The potentially good news is that their efforts can not only eliminate our dependency on foreign oil, but mitigate many other seemingly untenable problems ranging from pollution and job loss to terrorism. But their efforts have yet to coalesce around an agreed on approach that can overcome the obstacles blocking our way.

This book offers a way forward in the form of a guiding concept and a supporting strategy, distilled from a comprehensive investigation of our energy problems, the dangers they pose, and the potential solutions. Intended for the general public and government leaders, it describes how we got to this critical point in our history, what has held us back, and what each of us can do to stop polluting our planet, stop funding our enemies, regain our energy independence, and put modern civilization on a more secure footing.

S. L. Klein

Acknowledgements

Finding information on a subject today is far easier than it used to be, thanks to the Internet. Verifying the validity of the information, on the other hand, can sometimes be more difficult. I have acknowledged the importance of the Internet through the number of endnotes that cite Internet sources. Here I would like to single out for special attention the web sources that I found particularly useful in piecing together what is going on in energy. These sources range from our government agencies and our national laboratories to a number of important think tanks and universities doing research in alternative energy, oil, pollution, and climate change. Among so many on line, I'd like specifically to mention the Department of Energy (DOE) and its subsidiaries the Energy Efficiency and Renewable Energy (EERE) Program and Energy Information Administration (EIA); the Environmental Protection Agency (EPA); Sandia Laboratory; the National Renewable Energy Lab (NREL); the Oak Ridge National Lab (ORNL); the National Aeronautics and Space Administration (NASA); the Jet Propulsion Laboratory (JPL); the Air Force Research Lab (AFRL); the Electric Power Research Institute (EPRI); the California Energy Commission; the California Air Resources Board; the California EPA; Gibson Consulting; the International Center for Technology Assessment (CTA); the National Defense Council Foundation (NDCF); the Institute for the Analysis of Global Security (IAGS); and our leading universities: among them, MIT; Harvard; Stanford; the Universities of California; and the California Institute of Technology, which runs JPL. I would also like to acknowledge the information I have drawn from the numerous groups working to make alternative energy a reality. These include the Hydrogen Business Councils, Solar Energy International, and numerous other green organizations across the world.

Besides organizations, I would like to acknowledge what I have learned from our fellow citizens and citizens of the world who are inventing alternative technology.

More personally, I owe a great debt to Robert Klein for many discussions on technology, for reviewing technical portions of my manuscript and for acting as chief critic and editor. I've benefited greatly from his insights and suggestions. I want to thank Hank Wedaa for reading portions of this book and commenting and for giving me access to his large library on alternative energy. I would also like to thank my colleague Marcel Schoppers for reading and commenting on portions of this book. Finally, I would like to acknowledge Derek Jensen for his Twin Towers memorial photograph, which I used on my book cover.

The contents of this book, however, remain my responsibility. They reflect my research and opinions and are not intended to represent the opinions of any other individual or institution.

PART ONE

The Age of Hydrocarbons

> Science invites us to let the facts in, even when they don't conform to our preconceptions. It counsels us to carry alternative hypotheses in our heads and see which best fits the facts... This kind of thinking is ... an essential tool for a democracy in an age of change.
> Carl Sagan, *The Demon-Haunted World: Science as a Candle in the Dark,* 1996

1
The Amazing Twentieth Century

The twentieth century was like no other before it. Even in the nineteenth century, most people still rode to work on horseback or in horse drawn carriages as people had done for centuries. As late as the 1890s only a relative few had had the opportunity to live in the future: to ride in a horseless carriage[1] or to turn on a light bulb in their own home.[2] Yet in the last two decades of the nineteenth century, the world had reached a tipping point. What had become available to only a few would soon be possessed by multitudes. What would have been received with awe and superstitious fear by previous generations would become commonplace. We were entering an age of marvels unrivaled even by the Renaissance. Two key inventions of the Industrial Revolution made this unparalleled technological and scientific advancement possible: the alternating current (AC) power grid and the internal combustion engine. Without them, it is hard to imagine the subsequent development and manufacture of anything we consider modern from planes to cell phones. Unfortunately, we did not just reinvent civilization with these inventions. We made it dependent on hydrocarbons. We created an Age of Hydrocarbons whose darker consequences were unforeseen by all but a few.

Even today with hindsight it can be hard to fully appreciate the danger we are in or the irony: what energizes modern civilization could end up destroying it. Still hindsight can be an ally. Knowing the history of how we developed our current infrastructure and how we extended our use of hydrocarbons throughout our society provides the foundation for understanding where we have been heading, where we could end up, and what choices we have in the matter.

The War of the Currents

The invention of the AC power gird was largely the work of a maverick, an eccentric genius who in his later years talked of creating death rays that could destroy objects hundreds of miles away. Such claims, which seemed farfetched at the time, were hardly less incredible than what he actually accomplished. Nikola Tesla immigrated to the United States from his native Croatia in his twenties, after having already invented a prototype of the AC induction motor.[3] He would go on to file 111 patents for inventions ranging from the Tesla transformer and the Tesla turbine to neon lights. He would design our first power grid; and he would experiment in the wireless transmission of energy, the basis of radio.[4]

So stupendous a record of achievement should have made Tesla rich as well as famous. It did neither. He was destined to die in debt and in relative obscurity, eclipsed in name by his archrival Thomas Edison and by the businessman and industrialist George Westinghouse Jr. who bought his major patents. But in the 1890s the tragic events of Tesla's later years were still as unfathomable as his sudden apparent good fortune. His first major invention, the AC generator, had gained the support of George Westinghouse Jr., despite some very bad publicity.

The ostensible source of the bad publicity was Harold Brown, a professor, would-be inventor, and determined opponent of AC electricity. Through a series of public lectures held on the East coast, Brown evangelized against the dangers of alternating current in a cruel but compelling way. He electrocuted dogs, cats, calves, and horses on stage. The electrocutions drew mesmerized crowds whose fascinated horror effectively underscored the more gentlemanly protestations of Thomas Edison. Edison's opposition to AC electricity, it should be added, was far from academic. He was promoting the rival technology, direct current (DC) as the "obvious, safe alternative."

Despite Brown's theatrical demonstrations and Edison's more restrained opposition, Westinghouse offered Tesla the unheard of amount of $60,000 for his AC patents,[5] 150 shares of stock in the Westinghouse Corporation, and a $2.50 royalty per horsepower of electrical capacity sold.[6] More than money, Westinghouse offered Tesla an alliance, without which it is doubtful that Tesla could have withstood Edison, especially after General Electric took Edison's side in what would be dubbed "the war of the currents."

The campaign against AC reached its zenith when, alarmed by Westinghouse's initial success in selling AC generators, Brown decided on a countermeasure so extreme and grisly that even Edison must have been shocked, though he supported it. After purchasing a Westinghouse AC generator, Brown persuaded public officials to use it in 1890 – on a human being, William Kemmler, a convicted murderer. The first deliberate electrocution in history, it was graphically described in the New York newspapers and quickly dubbed "Westinghousing."

The association could hardly have been gratifying to Westinghouse, but it had little effect. By 1893, after a head-to-head competition with General Electric, Westinghouse won a bid to illuminate the Chicago World's Fair. His success would change the course of history. Ironically, Westinghouse Corporation outbid its rival and won, thanks to the very difference between alternating and direct current that Harold Brown and Edison had condemned: alternating current's much higher voltage capability. Using the transformers that Tesla invented, engineers could easily step up AC voltage to extremely high levels, something that Edison could not do with DC. Stepping up voltage is like increasing pump pressure to better push water through a pipe. Because of AC's high voltage "pressure," Westinghouse Corporation required only a single central power generator from which to distribute AC electricity through relatively thin wires over a wide area. In contrast, General Electric, because of direct current's lower voltage, [7] required thick copper transmission wires and potentially several localized coal powered generators.[8]

The technical differences between AC and DC translated into a huge advantage in time and money for Westinghouse, even taking into account the relatively confined space of the Fair. Westinghouse was able to bid half the amount General Electric proposed and to deliver on time, despite a tight schedule.

The deadline Westinghouse had to meet was the public opening of the Fair's exhibition devoted to electricity on May 1, 1893. Up to that day, few knew what Westinghouse and Tesla had accomplished, but the country was about to find out. To mark the occasion, the President of the United States, Grover Cleveland, came down to Chicago. His presence that spring evening elated a crowd already rife with rumors that something strange and momentous was about to happen. The crowd was not disappointed. With the flip of a switch, President Cleveland turned night into day using AC current. It was a defining moment.

The war of the currents had ended in triumph for Tesla and Westinghouse, but the struggle for power had barely begun. General Electric and Westinghouse Corporation, sometimes in league, mostly in opposition, now entered a new race to create a highly centralized AC infrastructure throughout the United States. With the financial backing of wealthy investors like J. P. Morgan, Lord Rothschild, and General Electric, Westinghouse Corporation took the initial step by creating the world's first hydroelectric plant at Niagara Falls. The plant would provide electricity to the nearby city of Buffalo, New York, turning Buffalo briefly into America's showcase for AC electricity and a real world laboratory.[9]

The basic hydroelectric system built by Westinghouse Corporation from Tesla's AC patents and designs is illustrated in Figure 1. Only in the illustration, a dam is used to artificially create the high-pressure flow of water produced naturally at Niagara Falls.

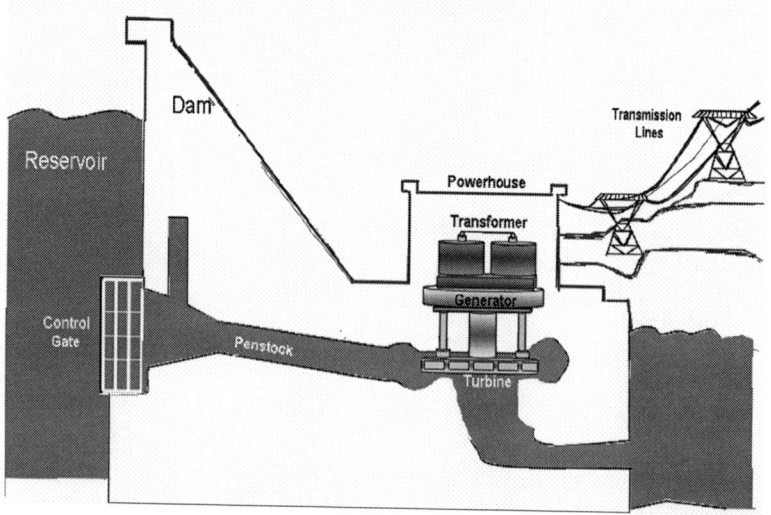

Figure 1. Generation of Hydroelectric Power[10]

Tesla's idea was to harness water to produce AC electrical energy. But how? The answer is one of the great scientific discoveries of all time. When magnets are rotated around stationary copper coils, an electric current is induced in the copper coils. Water creates the necessary force for rotation. Under the pressure of gravity, water flows through a turbine, forcing its huge blades to rotate. The rotating turbine blades turn shafts attached to them. The shafts rotate a circular plate above them. And the circular plate rotates giant magnets attached to it around stationary copper coils.[11] That is basically it.

Current induced by this generator (plate, magnets, copper coils) naturally alternates; that is, the electrical current is AC not DC. The alternating current is transferred through wires from the generator to transformers that sit above the generator inside a powerhouse. The transformers step up the electrical voltage, making it possible to push the electrical current across waiting transmission lines. Typically, electric current is pushed through three power lines, each of which transmits current at a different phase.[12] Why three lines? With three-phase power, one of the three phases is always at or near its peak at any given time. So the power output remains essentially even. One-phase and two-phase power output, in contrast, can fluctuate significantly, while four-phase power doesn't differ significantly enough in output from three-phase power to merit use.[13]

Though the principle of electrical generation through magnetic induction of current may appear simple, the initial costs were enormous. Financially drained by the Chicago World's Fair, by litigation in the battle over the new energy infrastructure, and by their takeovers of smaller companies like Thomas Edison's, the two electric giants Westinghouse Corporation and General Electric teetered on the brink of financial ruin. In this earlier era of robber barons, J. P. Morgan stepped forward, intent on controlling all hydroelectric power in the United States. Already an investor in the Niagara project, Morgan was in a position to discern the economic vulnerability of both Westinghouse Corporation and GE. He was determined to manipulate the stock market in order to starve them out and buy up Westinghouse's Tesla patents. Thanks to Tesla, he failed. Both Westinghouse Corporation and General Electric begged Tesla to give up his royalties' contract. Feeling responsible for the enormous expenses of the Chicago World's Fair and the Niagara project, Tesla agreed, allowing Westinghouse Corporation to recover. George Westinghouse died in 1914 having kept his dream. Years later, when Tesla fell victim to his own poor business decisions, no one stepped forward to help him, not even Westinghouse corporation, though no single individual was more responsible for Westinghouse's early prosperity and for that of the United States.

The Grid

Due to Tesla's sacrifices and genius and to Westinghouse's vision and support, on midnight November 16, 1896, a little over three years after the triumph in Chicago, the first power generated by Niagara Falls reached Buffalo, New York. Within five years the number of generators at Niagara Falls had reached ten, and the elements of the centralized grid system were taking shape. By 1900, power lines reached New York City. By 1908, AC powered Broadway's lights; a subway system transported people across the city; and American patents for electrical devices powered by the grid poured in.

What was taking shape was a monolithic power grid with networked regional power plants at its hubs and a number of companies controlling the grid under state and regional supervision. As shown in Figure 2, each power plant generates power mainly using water, refined oil, coal, natural gas, or nuclear power. Situated next to the power plant is a transmission substation that steps up the voltage so that huge, high voltage transmission lines can transfer the generated current long distances to power substations. Typical voltages for long distance current transmission range from 155,000-765,000 volts. The voltage is huge, but by comparison 765,000 volts is a bit under one percent of the voltage in a 1000 ft. lightning bolt.[14] Besides the transmission lines, there is a wire to deflect lightning. It is situated above and midway between the power lines. A typical maximum transmission distance for electricity is around 300 miles, at which point transformers intervene to step up voltage as needed. Eventually the power is transmitted to a power distribution substation whose transformers step the voltage down and distribute the current along power poles. Standard power line voltage is usually about 7,200 volts. By the time the current reaches a house, voltage is stepped down again to around 240 volts.

In older neighborhoods, this reduction in voltage is accomplished by a transformer drum, which hangs on the power pole near the destination site. The electric company runs two lines to your house, 180 degrees out of phase with one another. There is also a third line that runs from the power pole to the ground. Unlike the other two insulated lines that emerge from the transformer drum, this third line is bare. It is the ground wire, which sends unwanted electrical surges to the ground. If you live in a newer neighborhood, you will not see any power poles, much less transformer drums. The distribution lines that once hung on power poles will be underground, with above ground transformer boxes (rather than drums) the only visible evidence of the massive stationary power grid that controls the flow of electricity to our homes and workplaces.

THE AMAZING TWENTIETH CENTURY

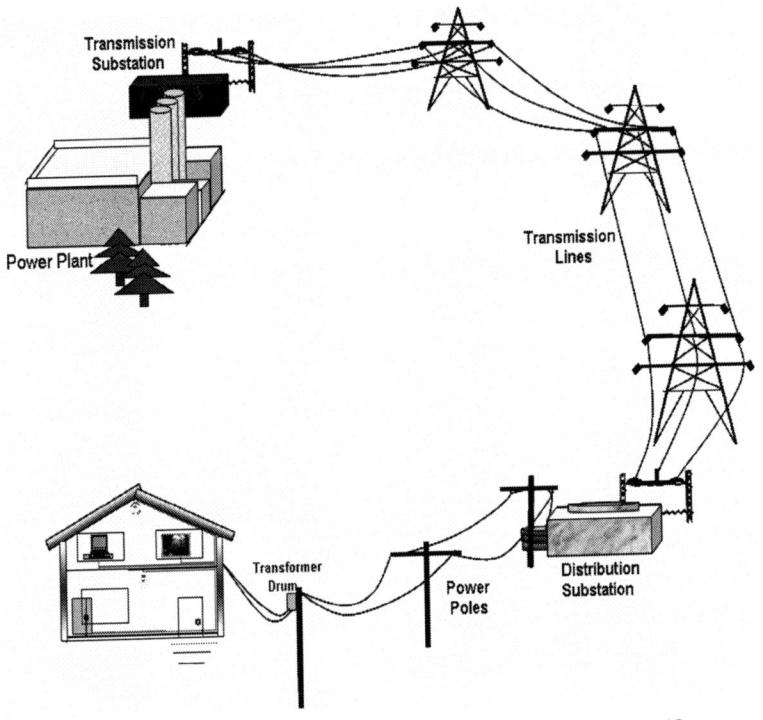

Figure 2. Basic Components of a Power Grid [15]

To protect the monolithic grid and isolate it from local problems, the power substations use circuit breakers and switches. These allow the substation to disconnect itself from the transmission grid or disconnect any of its distribution lines that might have failed. Redundancy allows power to be shunted from distribution lines that have failed to lines that can still carry power. In addition a bank of regulators helps maintain the voltage at proper levels.

The choice of an AC dominated grid was logical. It made economic sense. It cost less to set up than its DC alternative. It could be centralized. It worked reliably. It brought prosperity and a dramatic rise in the standard of living. What's more, it seemed miraculous to most of the public. If the modern world were to select its seven greatest wonders as the ancient world did, our grid would be on the list. So it's hardly surprising that few considered the implications of making huge numbers of people dependent on a power grid system radiating out from a relatively few vulnerable central plants. Nor is it surprising that few worried about the energy sources that would have to be used. Rather, the concern was to find enough energy to expand the grid, for it was quickly recognized that water was a limited option. If the grid were to expand, it would need a substitute source of energy to generate electricity. Coal, natural gas, and oil were the logical answers. Hydrocarbons made expansion of the grid across the country possible.[16] One of the few who foresaw any dangers in choosing hydrocarbons was Nicola Tesla.

A Car for the Great Multitude

Like the power grid, the internal combustion engine is ubiquitous. Almost all passenger cars in the U.S. have one. And like the spread of the power grid, the spread of the internal combustion engine has encouraged the development of an energy infrastructure that is highly centralized as well as highly dependent on hydrocarbons. But unlike the electrical power grid, the separate transportation network now depends largely on foreign imports of oil to function.

Comparable to the arteries and capillaries that bring blood from the heart to the local areas of the body, roads lead out from a relatively few central refineries to eventually reach the numerous service stations that dispense gasoline, diesel, oil, and other hydrocarbon products to waiting motorists.

Since 1980 the number of refineries in the U.S. has shrunk by about 50%, while their concentration has increased, further enhancing the trend toward centralization. Around 30% of our refineries can now be found in the southern states that border the Gulf of Mexico, an area subject to devastating hurricanes.

Looking back historically, our dependency on hydrocarbons for our transportation needs seems fated. Hydrocarbons were used to energize self-propelled vehicles from the start. The initial fuel of choice was coal. The preference relates to engine type. The first self-propelled vehicles had external rather than internal combustion engines. Coal was shoveled into a furnace beneath a boiler, external to the engine. The steam from the boiler expanded, pushing engine pistons that turned the wheels of a car with the aid of a primitive crankshaft.

Historians who consider steam-powered coal-fired road vehicles to be automobiles feel that Nicolas Cugnot (1725-1804),[17] a Parisian engineer and mechanic, was the inventor of the first automobile. Unlike modern cars, the self-propelled carriage Cugnot built in 1769 ran on three wheels. Looking more like an oversized child's tricycle than an automobile, it was by modern standards, dismally slow. It could go only 2½ miles per hour, which meant it was nearly but not completely useless. An engraving of a Cugnot shows the oversize tricycle being used for a solemn, stately transport of cannons.

Its apparent drawbacks did not prevent the eighteenth century public from becoming enchanted by a carriage that could move by itself. Despite its lumbering pace and disenchanting cost, it still seemed magical. Nor did its apparent impracticality deter would-be inventors. A few quickly realized that if the coal steam engine were scaled up in size, the weight, speed, and cost limitations could be overcome. By the 1870s, the most enterprising steam engine entrepreneurs had succeeded beyond reasonable expectation. They had invented a train that could surpass a speed of fifty miles per hour and carry over sixty passengers. This was not in the twentieth century but in the nineteenth century! Trains with external combustion engines continued to be produced well into the twentieth century. Quite a few steam-powered locomotives are still around today, relics of a bygone period.[18] Some more efficient steam cars were also invented, but as personal transportation they were soon eclipsed by two new automobiles, one evolutionary, the other revolutionary in design.

The revolutionary design came first. It made its debut around 1839. The electric carriage, as it was called, was an anomaly. It required no hydrocarbon fuel. It ran on electricity from batteries. But the inventor of the electric carriage, Robert Anderson, quickly encountered a problem similar to the one that had handicapped the steam powered external combustion engine. His electric carriage had a weight problem – not due to the engine itself this time, but due to the large batteries required to keep the engine running. Adding a further complication, the batteries severely limited the range that the vehicle could travel between charges.

Undaunted, electric car manufacturers began to spring up and find a genteel clientele willing to forego their horses for the electric car's quiet motor, comfort, and ease of use. Demand brought with it improvements and more models, one invented in America by Thomas Davenport around 1842.[19] By 1899, a Belgian, Camille Jenatzy, proved that a personal passenger electric car could overcome its speed problem. His single-man, rocket shaped electric car called the *Jamais Contente* reached a speed of over 105 kilometers per hour (about 65 miles per hour). By the 1890s electric cars in America were selling well – well enough to diminish the hopes of steam car enthusiasts, but not well enough to ward off a new rival whose origins were similar to that of the steam car but which had evolved its engine to operate not on external combustion but on internal combustion.

By some accounts, it took forty years from the advent of the electric carriage to come up with it,[20] but when the four-stroke internal combustion engine (ICE) came along, it transformed the automobile. The credit for inventing it usually goes to German engineer and entrepreneur, Nikolaus Otto, though there was at least one rival claimant. Otto's patent was challenged in France in 1886 by Alphonse Beau de Roaches. The dispute simmered for years, and can still spark arguments.

What made the invention of the four-stroke ICE a significant leap forward was not so much its internalization of the combustion engine as its use of gasoline. Because gasoline is an easily stored, relatively lightweight, concentrated source of energy, cars could be lighter, faster, and more powerful. And they could go much farther, before they had to be refueled, than an electric car could go between charges.

With its advantages, this new gasoline automobile ought to have overwhelmed its electric competitor from the start, but it didn't. It had serious drawbacks of its own from a customer's point of view. It had a complicated gear system - a forerunner of stick shifts - and a frustrating hand crank that was needed to get the car started. Still the gasoline car did prevail thanks to a series of landmark events that took place over roughly twenty years.

Among the most important events was the introduction of mass production around 1906. Henry Ford's assembly line made mass production of the gasoline automobile possible. It could have made mass production of the electric car possible too, but the electric car manufacturers chose not to mass-produce their cars – a strategic mistake that would cost them. By 1912, the electric roadster (one of the cheaper electric cars) cost $1,750 while the mass produced gasoline car sold for $650. But price was not the only difference. Important as it was, the difference in sales price would not have had such a dramatic effect if it had not been for trouble in the lamp business.

During the nineteenth century, the production of crude oil had grown into a huge business thanks to the world demand for a product that would seem unimportant today, kerosene lamps. Foremost among the companies that provided kerosene for lamps was Standard Oil. Because it had the infrastructure to produce kerosene, Standard Oil could produce gasoline as well. All Standard Oil needed was a little incentive, which it got unintentionally from Thomas Edison, an advocate of the electric car. The increasing popularity of Edison's incandescent light bulb began to devastate Standard Oil's customer, the kerosene lamp manufacturers. The kerosene lamp was becoming obsolete.

Growing almost desperate for another market, Standard Oil might have turned to the stationary power sector, but oil had a stiff competitor in coal. So it turned instead to the transportation sector where the situation was quite different. The gasoline internal combustion engine had triumphed over the coal burning external combustion engine, and now Ford had made mass production of gasoline automobiles possible. Transportation was wide open. All Standard Oil had to do was switch to gasoline production. But there was a little catch. By 1906, Standard Oil was running afoul of government anti-trust laws because of its monopolistic control of kerosene production in the United States. Never mind the growing obsolescence of the very industry for which kerosene was produced, Standard Oil was about to be broken up.

Making the situation even more nightmarish for Standard Oil was the discovery of more oil than anyone had imagined existed. The discoveries kept coming, starting with the Spindeltop oilfield in Texas in 1901 and the Masjid-I Suleiman field in Persia in 1908. Supplies flooded the market. Very quickly a glut drove down prices.

Meanwhile, by 1911 the anti-trust campaign against Standard Oil had succeeded. Standard Oil was broken up into Standard Oil of New Jersey and a number of other companies. After being weakened by the breakup, the American oil business might have been drastically scaled down due to the slacking demand for kerosene and the low price for oil, if it were not for the automobile. Instead, the breakup helped the oil companies. As the number of cars with internal combustion engines grew dramatically, the demand for oil grew too.

The companies that once formed Standard Oil saw their stocks suddenly double and triple in price as rich investors poured money into oil stock. Free to compete with one another, the oil companies were encouraged to bring increasing quantities of oil to market. The resulting oil boom drew even more investors while keeping the price of oil low.

The oil companies reduced the price of gasoline further through the use of thermal cracking and fractional distillation, techniques which improved the yield of gasoline from each barrel of crude oil. So the demand for gasoline cars increased not just because the gasoline cars were cheaper than the electric cars, but because gasoline was cheap.

The bad news for the electric car manufacturers was not over. In 1912, another major landmark event in automotive history helped the gasoline car defeat its rival. Charles Kettering, who would later become a General Motors vice president, unveiled a newly invented electric starter, an idea he got from the electric car. The new electric starter quickly displaced the gasoline automobile's old difficult-to-use hand crank, which had put off so many potential customers.

Then just two years later, the American oil business, after having stumbled onto a bonanza in the automobile, hit the mother lode with the start of World War I. The outbreak of war in 1914 illuminated the strategic military and political importance of internal combustion gasoline engines and consequently of gasoline. The engine's importance became clear early in the war when German forces began closing in on the beleaguered French outside Paris. On September 6, 1914, the quick thinking military governor of Paris, Joseph Gallieni, marshaled a fleet of Parisian taxis to meet the enemy – just in time. By using gasoline-run taxis to take troops to the front, Gallieni prevented the city from being besieged.[21] The U.S. government and governments across Europe quickly found other military uses for the internal combustion engine and gasoline. World War I saw England replace its antiquated coal fired, steam driven naval vessels with warships that had internal combustion engines. Coal was forsaken in favor of gasoline to improve maneuverability. And both warring sides began using surveillance and combat airplanes that required gasoline to fly.

With the growth in gasoline car sales and the stimulus of war, more oil companies began to emerge, merge, and redefine themselves – which for a very few like Winston Churchill began to awaken fears. Churchill's fears grew out of a famous merger. In Great Britain, the Samuel brothers had founded Shell Oil, named in honor of their father's seashell business. After a few years, they merged their company with Royal Dutch, allowing the Dutch to become the major shareholder and making England dependent on oil from a foreign owned company. Worried about the implications, Churchill urged the British government to take over a little British company called Anglo-Persian, the company that struck oil at Masjid-I Suleiman in Persia in 1908. In persuading Imperial Great Britain to take control of Anglo-Persian, Churchill had meant to ensure that oil from foreign countries would not pose an economic threat. But events did not work out as Churchill wished. The success of the Anglo-Persian Oil Company, now British Petroleum (BP), only encouraged the import of foreign oil to slake European and American thirst. France, which was largely shut out of the Middle East oil fields by American and British oilmen, acquired fresh resentments which were added to older ones. The investment in Anglo-Persian stimulated European and American rivalry. And the takeover of Anglo-Persian Oil Company by the British government completely failed in its objective when American oil companies began ceding control of the oil wells they built to the host countries. The gesture put immense pressure on other consuming nations to do the same.

So the politics of oil began to intrude, but only for a few. For the manufacturers of gasoline cars, little of this political maneuvering mattered. They were focused on prevailing in the contest with the electric car manufacturers. And they were about to get yet another assist. Congress passed legislation to construct more roads, allowing the car to escape the confines of the cities. This great escape made the differences in driving range, speed, and power between the electric and gasoline car even more glaring.

Thomas Edison, who by 1906 had recognized the adverse trend, had tried to intervene and save the electric car. As a devotee, he was determined to change the outcome by improving its battery. He recognized that the battery held the car back. But his attempts to create a longer lasting, more powerful car battery failed. His research yielded only modest improvement in battery performance. He abandoned the quest around 1910 and embraced the internal combustion engine as he eventually had to embrace AC electricity.

Without a compelling reason to continue the evolution of the electric car – pollution and oil dependency were not widely recognized as potential problems – by the 1920s the competition was essentially over.[22] Everywhere in peace and war, the internal combustion engine spread and with it the amount of oil consumed increased. Less than twenty years after the first Model T Ford was built, over 15,000,000 Model T Fords had been sold. By 1927, when the last Model T rolled off the assembly line, Ford and his competitors had realized Ford's dream of creating "a motor car for the great multitude."

A Revolution in Four Strokes

Since it achieved market dominance in the early twentieth century, the internal combustion engine (ICE) has retained its status as the overwhelming choice for passenger cars around the world, to our increasing detriment. To fully understand why the triumph of the ICE has become so problematic, you need to understand the ICE itself and its four-stroke operation. Understanding how the ICE functions will also further your understanding of the alternatives later.

The ICE propels an automobile forward using pressure changes within a confined space. The components of the internal combustion engine that make this possible are illustrated in Figure 3.

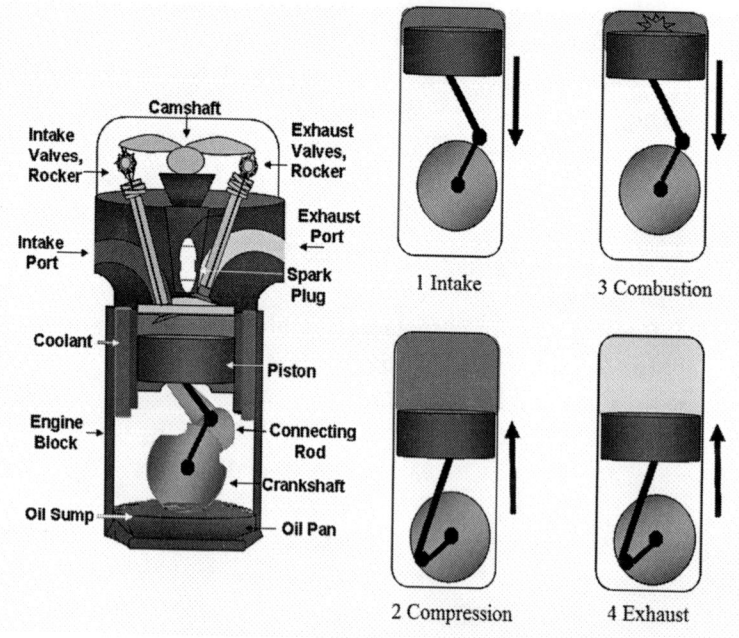

Figure 3. Internal Combustion Engine Cylinder[23]

Changes in pressure force the solid, cylindrical shaped piston to move up and down within the enclosing, hollow cylinder.

In order for the internal combustion engine to move a car, it must translate this up-down motion into rotational (circular) motion, which is used to turn the wheels. The vital translation is done through the crankshaft in four strokes: Intake, Compression, Combustion, and Exhaust.

The first stroke, the **intake stroke**, begins with the piston positioned at the top of its enclosing cylinder. At this stage in the engine's operation, the camshaft, which controls the engine valves, forces the intake valve open to let fuel into the cylinder through an intake port. The camshaft knows when it is time to open or close the intake valve thanks to a key principle built into the engine design: synchronization. The operation of the camshaft is coordinated with the piston and crankshaft using position. If you could keep your eye on both the piston and the fuel, you would not only see how the piston's position is synchronized with the behavior of the camshaft and the crankshaft, but you could also easily separate the four strokes from one another.

The gasoline-air mixture that enters the cylinder through the intake port comes from a carburetor or a fuel injector. The carburetor, a piece of very old technology, draws an amount of gasoline through fuel lines from the fuel tank, mixes it with air, and pushes the mixture into an engine cylinder when the intake valve opens. In modern fuel-injected engines, a more precise amount of gasoline is drawn from the fuel lines and injected into a cylinder as air is drawn in separately. The amount of fuel and air injected into the cylinder is determined with the use of a computer and sensors. In either case, as the piston begins to move downward under the weight of the fuel, it creates a space for the gasoline-air mixture flooding in. This space above the piston is called the combustion chamber.

During the next stroke, the **compression stroke**, the piston is forced upward by the motion of the crankshaft as shown in Figure 3. The effect is to compress the fuel-air mixture in the combustion chamber. Compression has the physical effect of heating and concentrating the mixture, which makes it more combustible.

In high-performance engines, which evolved from Otto's basic design, the air coming into an engine (cylinder) block through an intake port is already pressurized. Consequently, during the Compression Stroke even more air-fuel mixture can be squeezed into the combustion chamber. The increased compression through pressurization is called boost. Boost makes the downward thrust of the piston during the next stroke, the Combustion Stroke, more powerful.

There are two basic high-performance boosters: the turbocharger and the supercharger. Both use an air compressor to get boost. They differ mainly in where they get their power for the air compressor. The turbocharger air compressor gets its power from a small turbine attached to the exhaust pipe whereas the supercharger uses a mechanical connection to the engine to rotate the air compressor.

With the air-fuel mixture compressed, the engine is ready for an explosion, but it needs a spark. The spark is produced by the spark plug. The ignition of the fuel by the spark plug marks the start of the **combustion stroke.** When gasoline in the cylinder's combustion chamber explodes, the pressure of the explosion drives the piston down again as illustrated in Figure 3. The ignition of fuel inside the cylinder during this stroke makes the internal combustion engine a reciprocating (action-reaction) engine.

The spark plug that ignites the fuel during the combustion stroke receives its spark from a separate system, the ignition system, which has evolved over time. Today, many ignition systems coordinate their charge through an **ignition module** like the one shown in Figure 4.

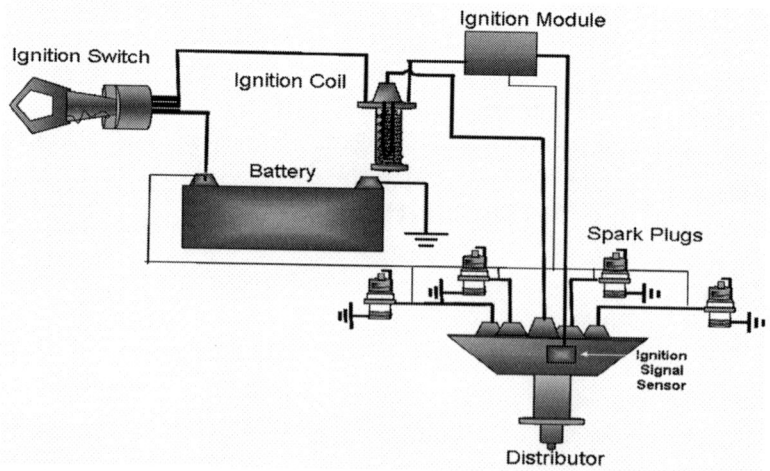

Figure 4. The Ignition System[24]

The timing is made more precise in modern ignition systems with the help of an **ignition signal sensor**. The ignition signal sensor is a position sensor that synchronizes the charge with the combustion stroke. Once given the signal, the ignition system produces a high-voltage electrical charge in an ignition coil. The charge is transmitted to each spark plug through the car's **distributor**.

After the Combustion Stroke, when the piston is again forced to the top of the cylinder by the crankshaft, spent fuel in the combustion chamber is squeezed out of the cylinder and pushed through the opened exhaust port into the tail pipe. This is the fourth and final stroke, the

exhaust stroke. From the tail pipe, the pollution is then ejected into the air.

In the illustration of the four-stroke engine (Fig. 3) only a single cylinder, piston, and spark plug were shown. Cars actually have multiple synchronized cylinders, usually four, six, or eight. These cylinders are arranged according to two basic configurations: **inline** and **V**. The different configurations have different advantages. Inline engines (usually with four cylinders) have the advantage of being cheaper because they are the least complicated, while V-engines (often with 6 or 8 cylinders) provide a smoother, more powerful ride.

Engines are also classified by cylinder volume. The difference between the minimum and maximum volume in the combustion chamber is called the **displacement** and is measured in liters or cubic centimeters (cc). One thousand cubic centimeters equals one liter, which is equivalent to roughly a quart.[25] Most conventional (internal combustion) car engines have displacements between 1.5 liter (1,500 cc) and 5.0 liters (5,000 cc). Displacement, because it is a gauge of how much fuel-air mixture an engine can hold, is also an indicator of how much gasoline a car consumes. The rate of consumption in turn affects the power of the conventional car.

As a throwback to the days of the horse and the horse drawn carriage, car power is measured as horsepower.[26] The name is not simply quaint. It is quite literally a measure of power based on the pulling strength of one horse. Horsepower is often expressed in terms of the weight that can be raised to a given height within a given time. For example, one horsepower may be defined as the ability to raise 330 pounds of coal or some other substance 100 feet in a minute.[27] Today the power of most cars exceeds the power of 100 horses.[28]

Just how much power modern cars possess can be measured using a dynamometer. The dynamometer places a load on the engine and measures the amount of power the engine can produce against the load. The result is translated into horsepower.

No other design feature of the Otto engine has been the subject of as much attention as engine power. Every possible means of tweaking the engine and its peripherals has been explored to increase power. The supercharger and turbocharger are just two devices used to make cars more "sexy," that is to say, more powerful. Generally, the more powerful the engine, the greater the amount of gasoline the car requires, the more pollution it expels, and the fewer miles per gallon it gets.

Not all of the improvements in the internal combustion engine were intended to improve power, however. Some were directed at solving functional problems. To ensure the smooth movement of the piston and the crankshaft, a circulating oil lubrication system was developed. As a result, lubricating oil became indispensable for operating an internal combustion engine. In most cars, the oil you put into your car goes to an oil pan such as the one in Figure 3. It is sucked out of the oil pan by an oil pump, run through an oil filter to remove grit, and then squirted under pressure onto the crankshaft rod bearings and the cylinder walls. The oil trickles down into the oil pan - the sump below the crankshaft - where it is collected and recycled.

To ensure that power is not lost through oil-gas leakage, pistons were girded with rings. The rings provide a sliding seal between the outer edge of the piston and the inner edge of the cylinder that confines it. The seal prevents the fuel-air mixture in the combustion chamber from leaking into the sump during compression and combustion. The seal has the additional advantage of keeping oil in the sump from leaking up into the combustion chamber where it would be burnt and lost. Most cars burn oil because the engine is old and the rings no longer seal properly. A sign of a faulty seal is the need to add a quart of oil every 1,000 miles or so. Another sign is a cloud of black or black and white particulates that streams from the tailpipe, especially when the car accelerates.

To ensure that an engine does not overheat, a cooling system was added. As the name suggests, an internal combustion engine operates at relatively high temperatures, high enough in fact that the engine is in danger of seizing. Seizure occurs when the metal gets hot enough for the piston to weld to the cylinder. If an engine seizes, it becomes useless. To keep the engine at a temperature below the seizure point, **coolant** is circulated. In most cars, this cooling system consists of the radiator and water pump. Coolant circulates through passages around the cylinders, travels through the radiator to be cooled off, and re-circulates around the cylinders.

For the modern four-stroke internal combustion engine to operate, it requires one other major addition to the engine, a way of getting started. "Turning over the engine," as it used to be called, means literally turning the crankshaft to initiate the four-stroke motion. That action was easier to understand when it was manual. You may have seen someone using the old hand crank method to manually turn over an engine in a faded silent movie or in a documentary on automobiles. Today, a sophisticated starter

THE AMAZING TWENTIETH CENTURY

system built on that of Kettering uses electricity to do the job automatically.

Turning over a cold engine requires a significant amount of current. Electricity transferred to the starter must overcome the friction caused by the piston rings. The starter system, like the spark plug ignition system, gets the electricity it needs to perform this operation from yet another system, the electrical system. The electrical system consists of a rechargeable **battery** that stores and provides the system with electricity, the **alternator** that generates electricity and transmits it to the battery for storage and later use, and a **starter solenoid**, a large electronic switch that can handle the high current generated when the ignition switch is turned on. The alternator is spun by a belt connected to the engine crankshaft in order to generate electricity to recharge the battery.

Over time, this electric system has grown increasingly more complex as it has taken on additional tasks. It has evolved to provide electricity to the radio, CD or DVD, the headlights, the windshield wipers, power windows, power seats, and onboard computers.

A Toxic Waste Dump on Wheels

As a feat of engineering, the four-stroke internal combustion engine (ICE) is a marvel. But despite its utility and its remarkable engineering, the ICE has serious flaws, which a burgeoning population can ill afford. It can productively convert only about 20-30% of the gasoline it consumes into mechanical energy. It wastes huge amounts of energy in the form of heat during operation, a major reason for its relatively low efficiency. More worrisome, it loses a significant amount of energy as unburned gasoline which, along with the products of combustion, is a major source of atmospheric pollution. Pollution and waste do not end with gasoline consumption, however. The ICE reeks of gasoline additives and detergents, lubricating oil, coolants such as ethylene glycol, transmission fluid, and steering fluid; all poisons. In short, the ICE is a toxic waste dump on wheels.

The Past that Wasn't

It took around 34 years (1893-1927) of increasingly accelerated growth to construct our modern energy infrastructure once the tipping point had been reached in 1893. Looking back, this infrastructure appears to have been built thanks to a series of interconnected events. Without Tesla and Westinghouse, the development of a power grid might have been delayed for decades. Without Ford's development of assembly lines and the use of interchangeable parts, the automobile could not have been mass-produced and would have remained out of reach for most people. Without a large, pre-existing, already centralized oil industry catering to a well-developed market, kerosene lamps,[29] no infrastructure would have pre-existed to produce gasoline. Without the invention of the light bulb, which threatened the kerosene lamp business, the oil companies might have been more reluctant to so completely switch markets. Without the discovery of large quantities of oil at home and abroad, limited supplies of gasoline would have inhibited sales of gasoline cars. Without World War I and the economic boom that followed, our government might not have been spurred to invest public funds in internal combustion engines and in roads, and the car might have remained pent up inside our cities much longer. Without all of these specific events coming together, the centralized power grid and separate centralized oil refinery system we developed might not have come about until decades later.

Still, though unique in specific details and spurred on by unique events, the development of our current energy infrastructure follows a generic pattern of market penetration. If you look into the history of the computer, the laser, the Internet, the CD and DVD, or the cell phone, you will see a similar pattern. Like the energy infrastructure, an individual product with some superior attributes reaches a tipping point, a point of maturity after which its commercial development and market acceptance begins. Companies arise with their version of the new technology and compete. Their first customers are usually relatively wealthy. These customers allow the product to get a foothold in the market. The product improves thanks to competition, and slowly at first the product's market share begins to climb. Prices decrease. Old competing technology perceived as inferior becomes obsolete. Some companies, new and old, fail, often pushed toward the brink by faulty business practices and by an inability to adapt. Amidst the resulting disruptions, the technology continues to spread, the prices continue to fall, and the penetration of the new technology deepens until it reaches a saturation point. The timeframe for major market saturation of a new product in its multiple variations may be no more than twenty years. Looking back, you generally find this generic pattern beneath the unique set of events that explain what happened.

Technology's characteristic mix - unique events underpinned by a general pattern - mimics biological evolution. Like biological evolution, the evolution of products and infrastructure follows a generic pattern but is profoundly influenced by the specific technology that arises. The new technology dictates what specific events make its acceptance possible. It profoundly affects the ecology of the pre-existing technological environment and it changes how we interact with our world.

Consider, for example, where we might be if the internal combustion engine had not been invented. What if we only had the electric car and external combustion (steam) engine as viable options well into the twentieth century? Undoubtedly, we still would have developed self-propelling vehicles, but trains, electric trolleys, and other forms of mass transportation might have become more predominant and for a longer period, changing how we move about. Similarly, if the AC power grid had not been invented or if DC electricity had prevailed, our stationary power infrastructure would still exist, but it would look different. The original transmission limitations of DC electricity would have made energy generation more localized and more expensive. At the same time, the need for more localized coal plants to generate electricity would probably have made pollution of greater concern earlier.

Unlike biological evolution, however, product and infrastructural evolution are controlled by the technologists who choose what new product "species" to introduce and by the public which chooses what succeeds. Usually the public prefers a product because it is more economical and superior to its competition in what it does. This is the usual rule of survival of the fittest in the marketplace. But we can be wrong in our choices – terribly wrong. We can choose a technology to build on, as we have done with our hydrocarbon infrastructure, that we discover belatedly can do serious, widespread harm.

Making a Course Correction

What do we do then? It would seem that we should use technology to correct technology, and that we should use the marketplace to introduce the correction. But how? Interactions within markets are complex, too complex for governments to control arbitrarily. Governments that have tried, generally enhanced the tendency of the market to become monopolistic and abusive. This tendency not only impedes growth and improvement, it is destructive and self-destructive. Outside U.S.

government control, partial monopolistic control of oil by cartels made the 1973 embargo possible. And unfettered monopolistic control of the marketplace caused the economy of the former Soviet Union to stagnate. The unfeasible five-year plans in the Soviet Union were ruinous, especially in agriculture and were possible only because the totalitarian Soviet leadership could impose them on its people against their will and against common sense.

Fortunately, there is another approach. Using our knowledge of the past, we could take advantage of the market's natural mechanisms and patterns to push it in a specific direction. That idea is part of the roadmap for change taken up later in Part Three, "The Age of Transition."

2

Pillars of Society

It is difficult to imagine just how great a quantity of hydrocarbons we consume. It is like comprehending the national debt. The consumption of oil alone is staggering. If the amount of oil we consume in just one day in the U.S. were stored in one gallon cans, the cans set side by side would encircle the Earth at the equator approximately six times.[1] How much we use is one indicator of its impact. What we use it for is another.

We have already seen how the use of hydrocarbons grew to impact the stationary power and transportation sectors thanks to two key inventions of the Industrial Revolution, the AC grid and the internal combustion engine. The power of hydrocarbons to transform our world didn't stop there. Thanks to their use in the heavy manufacturing sector, hydrocarbons mold the manmade material world around us from the infrastructure of our cities to the gadgets we carry around with us everyday.

The use of hydrocarbons in the heavy manufacturing sector did not, of course, begin in the twentieth century or even in the nineteenth century. Charcoal, a hydrocarbon, was used in the manufacture of tools and weapons thousands of years ago, during the Iron Age. But our enormous dependency on coal and oil in heavy manufacturing is much more recent. It began during the nineteenth century Industrial Revolution when we crossed a major threshold. Reaching that threshold contributed to the rise of darker, danker, unhealthier, more polluted cities than the European continent had seen since the Middle Ages.

What caused us to cross this threshold was a transformation in manufacturing processes epitomized by the Bessemer process.[2] First established around 1846 by Sir Henry Bessemer, this process was perfected over the next three decades. The Bessemer process is an innovative method for melting iron ore so it is not simply soft and slightly bubbly, but so it can flow like a liquid. In order to melt, iron ore must be subjected to temperatures from 2500-2900°F.[3] At the heart of the Bessemer process is a converter that generates these high temperatures. The converter is a large, pear-shaped furnace with holes at the bottom that permit the injection of compressed air.

Like many new inventions, this converter had an initial drawback that made it unworkable. While it made fire hot enough to melt iron, the compressed air it introduced removed too much carbon in the form of carbon dioxide from the melting metal. Without sufficient carbon, neither wrought iron nor steel can be produced. Recognizing the problem, another inventor Robert Mushet found a way to remove the overabundance of oxygen. He tied up the oxygen in manganese oxide, leaving less available to bond with carbon. His combination of manganese, iron, and carbon came to be called *spiegeleisen*, literally "mirror iron."

There was one other obstacle to overcome before the Bessemer process could be used as envisioned. Someone had to find an effective means of removing phosphorus from iron. Phosphorus makes steel extremely brittle.[4] In 1876, Sidney Gilchrist Thomas came up with the answer. He discovered that adding limestone to the converter drew the phosphorus out of the iron. His choice of limestone was logical. Limestone had been used since the Iron Age to improve the quality of iron tools – but now the reason it had been needed was understood. This discovery eliminated the final obstacle.

The Bessemer process that emerged with these improvements draws silicates and some other impurities out of the iron ore under intense heat. Most of these impurities flow to the bottom of the blast furnace while iron and around 4-5% carbon from coke collect on top. The leached out impurities at the bottom are called slag. Periodically, the liquefied mixture of iron, carbon, and some remaining impurities are siphoned off by the manufacturer and allowed to cool. The mixture when cooled is called pig iron, a somewhat pejorative term that denotes its hard and brittle characteristics. Further liquefaction of pig iron, chemical treatment, cooling, and beating remove enough of the remaining impurities for steel to be produced. The main impurity left in steel is a thoroughly mixed concentration of 0.2-1.5% carbon, a small amount that is essential.

With this technical triumph, more steel could be produced in a year than had been produced in the previous century. Because of the increase in steel manufacturing, the demand for charcoal – which came from thoroughly charred wood - exceeded supplies. A substitute was needed. The substitute found was coal coke. The use of coke in the manufacturing sector shot up exponentially. And with it, coal particulates began to darken the skylines of manufacturing cities across Europe and America while the forests gained some relief.

Today, steel plant owners still use a variation of the Bessemer process to which they add electricity (mainly generated using coal). Electricity is also now used to manufacture steel from scrap metal, increasing the role of hydrocarbons in manufacturing even more. Because of the Bessemer process and others that followed, steel became so abundant that it could be used as the material framework for our tallest buildings and could encase our smallest gadgets. It has become quite literally a pillar of modern societies.

As with the production of steel, the production of huge amounts of aluminum only became possible during the Industrial Revolution. Like iron ore, aluminum compounds had been used in manufacturing for many centuries. More than 7,000 years ago, Persian potters made pitchers and bowls from clay containing aluminum oxide. But aluminum was not shown to be a metal separate from the compounds in which it naturally occurs until 1808 when Sir Humphry Davy proved it existed in its own right and named it. It wasn't separated from its compound until 1821 when the Danish physicist Hans Christian Oersted managed to produce a minute amount, considered more precious than rubies. For a long time afterwards, aluminum remained too costly to produce commercially.

In fact, it was not possible to produce more than an extremely small amount at a time until the 1880s when Charles Martin Hall made a momentous discovery. He found the secret to extracting aluminum in large quantities at low cost was electricity.[5] If an electric current is passed through a molten alumina compound, aluminum is separated out.[6] Hall was not the only one to make this discovery. Around the same time, an Austrian chemist Karl Joseph Bayer independently developed a similar process.[7] To capitalize on his discovery, Hall and financier Alfred E. Hunt founded the Pittsburgh Reduction Company now known as the Aluminum Company of America (ALCOA) in 1888. In 1889, Hall patented the process. Like steel, once it could be produced in mass quantity, aluminum became indispensable to modern civilization.

Much the same pattern of dependency on hydrocarbons developed with manmade plastics. Like the manufacture of steel and aluminum, the manufacture of plastics requires large quantities of coal, natural gas, or oil. Unlike the metals, however, plastics are themselves composed of fossil hydrocarbons, at least today. That was not initially the case. The first manmade plastic invented by Alexander Parkes was made from a celluloid derivative, a plant sugar. This plastic called Parkesene had the characteristics with which we are all familiar, characteristics that awed the crowds who first saw it at the 1862 International Exhibition in London. It was rather transparent though not as transparent as clear glass. It could be molded like steel but at a fraction of the temperature, and it would retain its shape after cooling while remaining much more pliable than steel.[8]

There was one problem. It was too expensive. Before plastic could become the next word in future wealth and prosperity,[9] someone had to figure out how to make it more cheaply. And of course someone did, several people, in fact. Those who deliberately set out to find cheaper substitutes for the celluloid derivative and those who found them serendipitously didn't have far to look. The monomers of hydrocarbons are related to the Parkesene cellulose derivative and work incredibly well as substitutes. By using several oil processing techniques including thermal cracking, unification, and alteration, researchers found they could separate out these monomers from the hydrocarbon stew and link them together in long chains called polymers.[10] The results were polyethylene, polypropylene, polystyrene, and a myriad of other plastics, all unknown to nature.

Through molds, additives and plasticizers, researchers also discovered that the polymers' characteristics could be enhanced so that they resisted the degrading effects of light, heat, and bacteria. Polymers could be squeezed, tugged, tucked, melted, colored, and grown into so many products that it's hard to find a computer, cell phone, refrigerator, car, train, plane, stick of furniture, dress, pair of shoes, lamp, or filing cabinet without them. Along the way, the name Parkesene was replaced by the name plastic, perhaps because Parkesene sounded too much like an Epoch in the geological timeline, stuffed somewhere between Pleistocene and Paleocene.

So indispensable have plastics become, so amazingly versatile and diverse, so cheap and disposable that we would be hard pressed to find substitutes. And so widespread are plastics that today scientists concerned with environmental degradation are unlikely to find a single large body of water devoid of plastic garbage or a single beach in the world that is not strewn with minute to sizeable pieces of plastic mixed in the sands.

Like the widespread use of gasoline engines, the AC power grid, steel, aluminum, and wrought iron, the ubiquity of plastics has only deepened our dependency on hydrocarbons.

3

A Place at the Table

Hydrocarbons not only transformed our transportation, stationary power, and manufacturing sectors, they revolutionized agriculture. Through fertilizers, hydrocarbons promote phenomenal plant growth. As pesticides, they guard our crops against insect predation. Without them agriculture would be in deep trouble.

That fossil hydrocarbons are used in the manufacture of fertilizers may not be immediately obvious, especially to the amateur gardener who uses natural animal waste as fertilizer. Yet nearly all fertilizer manufacturers use hydrocarbons to produce the most essential plant nutrient, nitrogen. The best-known artificial fertilizer manufacturing process is the Haber-Bosch process first developed in Germany before World War I by Fritz Haber and Karl Bosch. The process generates a reaction between ammonia, hydrogen, and nitrogen gases. The hydrogen gas is derived from hydrocarbons, usually from natural gas. But oil and coal can both be substituted to produce hydrogen. The process also requires high temperatures, high pressure, and a catalyst (oxides of iron) to produce the end product, ammonium nitrate fertilizer.

As a side note, ammonium nitrates are not used just for fertilizer. Nitrates can be used to make powerful explosives. By 1916 the German war effort might well have been stymied by a dearth of nitrates if it had not been for the discovery of the Haber-Bosch process. Faced with the same problem but without access to the new process, the allies found their war effort seriously threatened. They were able to continue because they got access to a natural deposit of Chilean nitrates, but there was a catch. Their dependency on the Chilean nitrates subjected them to price hikes and supply restrictions from a cartel that arose to control the supply.

Because traditional organic fertilizers such as manure, stubble, and compost, cannot supply sufficient quantities of nitrogen to maintain large-scale agriculture, farmers have had to turn increasingly to manmade nitrate fertilizers to maintain high crop yields. These fertilizers save millions from starvation, but they are a mixed blessing. Especially since 1950, they have promoted human population growth at incredible rates,[1] in contradiction to Thomas Robert Malthus's prediction in the nineteenth century that human populations would be held in check by limited food production. He assumed that food production could not keep up with an exponential growth in population.[2] It has. The growth of manmade fertilizers mirrors the growth in population with one exception. There was a slight dip in ammonium nitrate fertilizer production in the 1990s after the collapse of the former Soviet Union.[3]

Like the use of other hydrocarbon products, the use of fertilizers is not equally distributed. Africa continues to have problems obtaining sufficient fertilizers to grow food. In contrast, by far the greatest growth in the use of manmade fertilizers today is in China and India, which together constitute one third of the world's population.

The International Fertilizer Association estimates that less than 2% of the world's energy consumption is due to the manufacture of fertilizers,[4] but without these fertilizers, our first world populations could not be sustained. The dependency on manmade fertilizers continues to grow ominously, abetting the growth in population.

As with fertilizers, the use of hydrocarbon-derived pesticides has grown significantly since the 1950s, mirroring population growth. Pesticides came into their own after World War II with the introduction of DDT, BHC,[5] dieldrin, and endrin. The manufacture of hydrocarbon-based pesticides has its roots in the refinement of oil and the manufacture of plastics, but uses additional steps such as esterification. Esterification is a chemical reaction in which two chemicals form an ester. An ester is an organic compound of an organic acid and an alcohol. The process requires a catalyst.

Relative to artificial fertilizers, pesticides have gotten much more negative press, thanks in part to the continuing influence of Rachel Carson's book *Silent Spring* (1962), which described the deadly effects of DDT, especially on bird populations. But insecticides really deserve mixed reviews. Without them, it is difficult to combat diseases such as malaria that kill millions or to protect crops from insect predation. R.D. Knutson, a professor emeritus at Texas A&M University, estimated that a pesticide ban in the U.S., for example, would decrease yearly supplies of corn, wheat, and soybeans by 73%.[6] On the other hand, powerful insecticides like DDT are indiscriminant killers that can kill hundreds of insect species as well as harm many other species. Though they might protect a fruit-bearing tree from insect predators, for instance, they could also harm beneficial insects like the honey bee that pollinate it. Some might argue that organic farming and the use of natural enemies to combat crop pests could eliminate insect infestation without recourse to pesticides. But these countermeasures are currently too limited in scope and effect to act as substitutes. So despite the damage they may do, pesticides like fertilizers are assured a place at the table because the consequences of not using them are, so far, more frightening than the consequences of using them.

4

Dependency and Paradox

Hydrocarbons are the stored energy of creatures that died millions of years ago. They are the remains of other lives that we have resurrected like ghostly servants to do our bidding. They have transformed our lives, making possible a lifestyle unlike any other in recorded history. They wash our dishes and clean our clothes. They ferry us to work and to play. They guard our food supply and nurture its growth. They ensure the transport of vital goods and services. They make possible the skyscrapers that crowd the skylines of our big cities. They touch nearly everything around us, nearly everything we consider modern. Without them we could not today sustain civilization as we know it. Unfortunately, these compounds that make modern civilization possible also poison the air we breathe, the food we eat, and the water we drink, and that is only part of the problem. The darker implications of our alteration, combustion, and consumption of hydrocarbons are more fully explored in the final chapters of Part One. What is important to grasp at this point is the true scope of our dependency, which many of us have never fully considered, much less come to grips with.

5

Oil and the Third Wave of Totalitarianism

The twentieth century saw an unprecedented spread of Western democracy in defiance of historical precedence. Absolute monarchy, imperial rule, strongman dictatorships, and other forms of one-man and oligarchic dictatorship have predominated for centuries in the West as well as in the East. The ideals of Western democracy – basic human rights, separation of church and state, equality under the rule of secular law, freedom of the press, freedom of religion, freedom of expression, and representative government – are historically rare. When they do arise and take hold, they provoke fierce opposition.[1] To its enemies, Western democracy inspires fear and hatred because it permits a self-directed life while instilling a belief in the collective wisdom of the masses. Nothing could be more inimical to authoritarian or totalitarian thinkers, religious or secular – though to some of their misled followers their contempt for the unfettered mind and for democratic decision making may be initially obscured. To authoritarian or totalitarian thinkers, democratic values are not simply alien concepts; they are a threat to the idealized social order these thinkers envision.

The despots and demagogues who resisted the growth of democracy in the twentieth century often claimed that history was on their side. Twice they tried to assert totalitarian world rule with potential success, and twice they were put down, though not entirely.[2] In each case, most historians who lived in democratic states gave little or no credit to the role oil played in democratic success.[3] However, the totalitarian and democratic leaders who fought the twentieth century's wars, hot and cold, were acutely aware of how important oil was.

During World War II, Nazi U-Boats were sent into the North Atlantic specifically to torpedo oil tankers bound for England. Until the number of U-Boats was significantly reduced, the strategy worked.

England was in such distress that it was threatened with starvation. Such was the crucial role of fuel from oil in transporting food to an island that could not grow enough of its own.

General George Patton was prevented from invading Germany and possibly concluding the war in Europe because of want of fuel. His tanks were literally stopped in their tracks. The bombing of Germany's synthetic fuel industry doomed the *Luftwaffe* just when it had gained a tactical advantage from its introduction of the jet fighter. In both the Pacific and the Atlantic theaters, the success of naval battles frequently hinged on whether or not enough oil was available. The need to secure oil supplies is one of the reasons, if not the main reason, that oil-poor Japan embarked on the aggressive conquests that led to its alliance with Germany and its war with the United States.

The post World War II Cold War struggle that pitted the Soviet Union against the United States was, in the Middle Eastern theater, primarily a struggle over oil. The denial of vital resources to the West was one of the Soviet Union's strategies in its struggle to defeat America and establish communism across the world. That strategy was hampered by America's ability to produce most of what it needed including oil, a situation that lasted almost to the end of the Cold War.

As we begin the twenty-first century, we are once again engaged in a battle with a global totalitarian ideology. What is different now, from an energy perspective, is that we are no longer self-sufficient in the cheap, easily accessible oil on which our economy has depended in peace and in war. The decline in domestic oil production, which began in earnest in the 1970s, has increased our dependency on oil imports. Europe has been even more dependent on oil imports than we have been.

Given its importance, the idea that the Western democracies should allow themselves to become critically dependant on some of the world's most despotic, hostile, and unstable regimes for so vital a commodity as oil ought to be considered madness. But neither our historians nor our media have made sufficient effort to warn the public, much less to demand change, while the public continues a habit of waste that increases our dependency.

Nor has sufficient attention been given to the dangerous aid we are giving to the enemies of Western democracy through our petroleum dependency. Instead, faced with a third global wave of totalitarianism in a century, many seem to be falling back into the same mental traps of denial, fantasy, and self-flagellation that characterized the West before World War II.

The Gathering Storm: Parallels with the 1930s

When the Nazis began rearming Germany in the 1930s during the period Winston Churchill characterized as the "gathering storm," the democratic countries of Europe failed to respond. Europeans resisted calls for self-protection, and intellectuals across Western Europe expressed a strong desire for accommodation aimed at appeasing the Nazis. Few who called for appeasement appear to have read *Mein Kampf,* which clearly describes what Hitler had in mind. If they had, they might have realized appeasement could not work because Hitler was not interested. On the contrary, he took expressions of appeasement as signs of weakness and publicly gloated at the obvious fear they suggested. Instead of moderating his rhetoric, he intensified it, fanning a sense of humiliation, resentment, and pride that increased as Germany grew stronger. When Hitler marched into the Rhineland unopposed in 1936, it was to great applause among a growing constituency. Similarly today, Bin Laden, in a propaganda campaign against Western democratic values, plays on a sense of humiliation, resentment, and pride among Muslims to gain recruits for his Jihad, while Western countries do little intellectually to counter his claims. Extreme Islamic fundamentalists and dictatorial governments have the mental battlefield virtually to themselves in the Middle East.

What should be particularly unsettling to Western intellectuals is the fact that Bin Laden's ideas don't come exclusively from extreme Islamic writers. Bin Laden's group has been strongly influenced by Nazi and Communist ideologies and methods, something not sufficiently acknowledged in our media. His worldview is consistent with the virulent anti-Semitism of the Nazis and the virulent anti-Capitalism of the Communists, and reflects a contemptuous perception of the Western democracies as weak, decadent, sexually depraved, and selfish. The myriad sex scandals and wanton violence reported nightly in America serve Bin Laden well in promoting this view of the West and of America in particular.

Because of Al Qaeda's stand against Capitalism and because of their anti-Semitism, Bin Laden's group has been winning some not-so-secret allies among extreme radicals and reactionaries in the West. In return for espousing extreme left and rightwing ideas, Al Qaeda gets its freshest supply of propaganda from them including over-the-top theories of U.S. government conspiracies that are passed off as indisputable facts. These include the assertion that the U.S. created the AIDS virus and is using it to destroy Africa; that the American government detonated an atomic bomb to cause the Christmas Tsunami in Indonesia; that FEMA set up a concentration camp for blacks in New Orleans after Hurricane Katrina hit; and that agents of the Bush administration destroyed the Twin Towers and damaged the Pentagon on September 11, 2001.

Although Bin Laden regularly uses the language of the Far Left in his characterizations of American democracy, he does deny one of their assertions. He now quite openly takes responsibility for 9-11. Lest anyone doubt it, he had Al Jazeera publish his claim of responsibility on its website. The claim was translated into English.

Many on the Far Left indirectly defend Al Qaeda by comparisons that are intended to shift attention to the U.S. The Far Left points out that Al Qaeda may say it wants to detonate an atomic bomb, but the U.S. actually destroyed cities using atomic bombs. Al Qaeda may want to use biological and chemical weapons against its enemies, but the U.S. actually developed an arsenal of biological and chemical weapons for that purpose, and so on. The intent is not even to set up a moral equivalency between the U.S. and Al Qaeda, but to portray the U.S. as the real threat. If this interpretation is taken to its logical extreme, Al Qaeda's terrorists become freedom fighters, and the U.S under its "Fascist" President is out to dominate the world.

These are potent arguments particularly when presented to people unfamiliar with any culture but their own. But what those who espouse this position fail to recognize is embodied in a simple question. Under what government would they really prefer to live: the one set up by the Taliban, which conformed to Al Qaeda's ideals or the one under which U.S. and Western European citizens live? Forget the difference in wealth. Consider the difference in values and constraints on power.

The Far Left in America has made a lot of money selling books proliferating anti-Americanism throughout the world. And if challenged by the American government, they have gone to court with their lawyers to condemn the violation of their rights. They usually win. Imagine if they had criticized the Taliban while living under its rule. Consider further what Al Qaeda would do today if it were the sole possessor of an arsenal of atomic bombs as the U.S. was just after World War II. Is there any doubt? Al Qaeda's thinking is sociopathic and largely without constraint. The ends – God's rule on Earth – justify the means or whatever it takes: enslavement, mafia style intimidation, torture and beheading of helpless captives, and indiscriminate mass murder. The key idea that makes all this acceptable to Al Qaeda's followers is simply, if the wrong person is killed, it won't matter, because God will sort things out in heaven. Try arguing with that logic.

The alliance between the most extreme of Western radicals and extreme Islamic fundamentalists recalls the attitude of the less extreme anti-war French in the 1930s and the pro-Stalinist Communists in America before World War II. The anti-war sentiment in France just before the outbreak of World War II is particularly telling in retrospect.

Initially, the French Left seemed to underestimate the growing power of the Nazis. Even after the Nazis' rearmament of Germany could not be denied, the French government led by Leon Blum, continued to resist any preparation to protect the country. Instead, Blum's government, which was a coalition of Communists and socialists, persisted in arguing that French rightwing imperialists and their big business allies were war mongering. The left wing leaders saw the rightwing motivation to arm as a desire for self-enrichment at the expense of the country. They warned that if the war mongers were heeded, France would be plunged into another devastating, useless war like World War I. Going even further, some of the socialist and Communist intellectuals adopted Nazi propaganda, which reinforced left wing thinking that unscrupulous Jewish financiers were foremost among the war mongers out for profit. When war came and France was drawn in, the French Left fractured. A few of its leaders actually joined the new Vichy government and swore allegiance to Petain and to Nazi Germany. Most, hunted by the Nazis and Hitler's Vichy allies, went underground and became part of the resistance. The French leader Leon Blum, a Jew, was taken to Auschwitz. Miraculously, he survived and returned to France after the war.

It is easy, with hindsight, to condemn the anti-war socialists in France for their actions, their illusions about the nature of the Nazi regime, and their pacifism. Given their experience of World War I, their fear of provoking another devastating war ought to be understandable. Many of the beliefs underlying their desire to accommodate Germany also made sense. Arms manufacturers in France did stand to profit from rearmament. And France had wanted revenge at the end of World War I.

France had prevailed over the opposition of Woodrow Wilson in its insistence that Germany be harshly treated. Many Frenchmen now felt France had been too harsh.

What the French government and its adherents failed to see was the bigger picture. Hitler drew upon his own personal sense of humiliation and anger, but revenge for Germany's World War I humiliation was only his starting point. It was an excuse for conquest. Too many people narrowed their focus to what seemed like justifiable anger on the part of the Germans and came to the wrong conclusion that Hitler could be appeased. Too often, primary sources like *Mein Kampf* are never read or if read are not believed – such is the power of wishful thinking. We continue to overlook, minimize, or deny the clear messages sent to us. So it is not surprising we put up so little in the way of an intellectual counter to the threat.

Worsening the position of the Western democracies was the behavior of Stalin, who felt threatened by the West and at the same time hoped to capitalize on a war between the Axis Powers and the West. His alliance of convenience with Hitler was a betrayal of ideals professed by Western left wing intellectuals in Europe and America, just as his "dictatorship of the proletariat" had been. Yet many left wing radicals went along with Stalin, perhaps because his stated objective was to sit back and let the Western democracies and Nazis decimate one another, after which he expected to step in.

Stalin succeeded in dividing the democratic West further, which made the West's response weaker and encouraged an aggression that eventually backfired on him. But before Hitler betrayed him, Stalin had already changed from his semi-neutral stance to a more aggressive one. He was quite willing to get involved in Hitler's unopposed aggression, and willing to reap some of its spoils. His actions foreshadowed what he would do in Poland and the rest of Eastern Europe after the war. While the Soviet Union remained an ally of the Nazis, Communist demonstrators in the streets of America opposed American intervention and characterized American opposition to Hitler on the grounds that it was imperialistic.[4]

Like the alliance of the 1930s, the seeming accord of the Far Left in Western societies today with Al Qaeda cannot hold. In step with Al Qaeda's denunciation of American democracy, Al Qaeda's followers in Europe have not been shy about proclaiming their ultimate intention for Europe, which is not to make it into a socialist paradise. They wish to make it into Dar Al-Islam, a realm of Islam, by ballot or Jihad. In a 2004 London interview Sheikh Omar Bakri Mohammed not only spoke of this goal but he suggested that he would personally see the day the black flag of Islam flies over 10 Downing Street.

The World's New Tinderbox

As the Middle East has grown wealthier, it has become more polarized and dangerous. Like the Balkans before World War I, the Middle East has become a tinderbox that could set the world on fire. It is into this highly unstable region, seething with resentments against America and Europe, that Western countries have poured huge amounts of money in order to get oil[5] despite the despotism of so many of its leaders and despite the ideological campaign being waged against the West, its liberal values, and ideals.

Oil money that hasn't been used by despots to develop weapons of mass destruction continues to be used for resplendent palaces, grandiose statues, and gigantic portraits of despotic leaders, despicable hallmarks of totalitarian regimes. Oil money has maintained the opulent lifestyles of warlords in a community of nations that derive their structure from tribal alliances and kinship. It has put weapons into the hands of combatants who adhere to an ancient law of revenge (*lex talionis*). It has contributed to frequent blood feuds that have shocked many of us by their ferocity and cruelty. But even more telling, it has paid for the religious schools, the madrassas.[6]

Religious Intolerance

A major recipient of Western oil money, Saudi Arabia, has directed huge amounts of money into building a network of conservative religious schools that spread the Kingdom's extreme and intolerant version of Islam based on Abd al-Wahhab teachings.[7] In these madrassas, young Muslim boys are taught Allah's ultimate plans for non-believers once Islam has conquered the world: death, conversion, or a form of subjugation called dhimmitude.

Dhimmitude is demanded in the Koran and increased in scope and magnitude by extreme Islamic fundamentalist philosophy, Sharia (Islamic law), and local custom. When Bin Laden issued the demand that Americans convert to Islam or face the consequences, it was not simply to mollify critics who blamed him of attacking without offering peace through conversion first. It was not just to justify whatever further atrocities he wished to commit against innocent civilians. It was in accord with an interpretation of selected passages in the Koran such as this one:

> "Fight those who believe not in Allah nor the Last Day, nor hold that forbidden which hath been forbidden, by Allah and His Messenger, nor acknowledge the religion of Truth, even if they are of the People of the Book, until they pay the Jizya with willing submission and feel themselves subdued." (Sura 9:29).

The religion of Truth is, of course, Islam; the Messenger, Mohammad; and the People of the Book are Jews, Christians and Muslims. Jews and Christians are People of the Book because part of the Old Testament, or an interpretation of it, is incorporated into the Koran, and because Jesus is recognized in the Koran as a prophet of Allah.

The command that a Jizya (tax) be paid by non-believers might first appear to be an enlightened policy, at least for the period in which it was first promulgated. It meant that the non-Muslim People of the Book who paid the tax were declared dhimmis or "protected." But within Islamic societies that abide by extreme fundamentalist forms of Islam, what dhimmitude has come to mean, contrary to what might be supposed from the word "protected," is that these infidels are subjected to severe restrictions on education and career, meant as reminders of their subservient, second-class status and "justified" humiliation. The "unprotected" people may expect to receive much worse treatment.

Needless to say, such teachings and laws are hardly compatible with equality under the law in secular Western democracies or with the universal declaration of human rights, far from it. Second class citizenship is not only mandated for all non-believers by religious extremists in the most conservative Muslim societies, it becomes a basis for further violations of human rights. Under Sharia (Islamic law) anyone believed to be in violation of dhimmitude's conditions is subject to death. Such a violation can be easily alleged and used to justify mass murder. The Turks have been accused of using a violation of dhimmitude as a pretext for the massacre of Armenian Christians in 1915, a massacre many have characterized as genocide.[8]

The practice of dhimmitude promoted in the conservative madrassas is akin to the practice of slavery, which has never entirely died out. On religious grounds, conservative Islamic countries have been the most reluctant to give up either. Shamefully, the Saudis did not renounce slavery until the early 1960s. In some Islamic states it has never really been renounced. A major reason is that slavery was an accepted practice in Antiquity, was described in the Old Testament, the New Testament, and the Koran and was never unequivocally condemned in any of these books. In the Koran, enslavement of infidels is an acceptable practice.

Together with slavery, extreme Islamic fundamentalists demand adherence to a set of cruel and unusual punishments specified in the Koran. These range from cutting off the hand of a thief to stoning an adulterer to death.

Not surprisingly, given dhimmitude, the cruel punishments demanded in the Koran, and the attitude toward enslavement, the idea that women might be afforded anything close to equal protection under the law is blatantly dismissed. Within Taliban Afghanistan (the best documented example of the extreme fundamentalist Islamic state Bin Laden has in mind for the world) young women were not even allowed to be educated, though women were allowed to beg in the streets. Women had to appear in public completely covered except for a grid of little pinholes, which was permitted so that they could decipher where they were going. Females of all ages were subjugated to males; wives to their husbands, mothers to their sons, sisters to their brothers. Courts were bound by Sharia law to give greater weight to the testimony of a male witness or defendant than to a female. Females, though less likely to break the laws, were more likely under Sharia than males to be found guilty in cases where the accuser is male. A male accused of a crime by a female was more likely to get off lightly no matter what the evidence might be.

In conservative Islamic states, women may even be punished for crimes committed by male relatives. In one village in neighboring Pakistan, a woman was reportedly sentenced by religious leaders to be gang raped when it was found that her brother had been seen in the company of an unchaperoned, unidentified girl who was not a close relative. The rape was meant to humiliate the family.[9] Often female rape victims are doubly victimized, first by the rapist and then by their own families who feel obligated to kill the rape victim to cleanse the family of the humiliation of unlawful intercourse. This act is called "honor" killing.

In parts of the Islamic world, among the devout, female circumcision is still practiced. It is often performed without anesthesia. This circumcision is performed to deny women the physical pleasure that might cause them to stray from the marriage bed. When they do stray, under Sharia law they may be stoned to death. Men rarely suffer such punishment. Why? It is believed that female adultery, if allowed, would completely destabilize the social order.

Supporting the low status of women are passages from the Koran. Many passages can easily be construed not only to justify female subservience, but also to justify physical violence against women. According to the Koran:

> "Men are in charge of women, because Allah both made the one of them to excel the other, and because they spend of their property for the support of women. So good women are the obedient, guarding in secret that which Allah hath guarded. As for those from whom ye fear rebellion, admonish them and banish them to beds apart, and scourge them. Then if they obey you, seek not a way against them. Lo! Allah is ever High, Exalted, Great." (Sura 4:34)

In the most conservative Islamic states, women are not even permitted to leave the house unless a male relative accompanies them. Their activities as children are restricted for fear that they may rupture their hymen, which might subject them to charges of adultery and would, in any case, make them technically unmarriageable. Women are not believed responsible enough to make their own decisions, and in some of the most conservative countries, education is denied them. The apparent reason is a fear that they will think for themselves and stray.

In the eyes of extreme Islamic fundamentalists, the economic and social freedom enjoyed by women in Western society is decadence; individualism is selfishness; representative government, self-reliance, and self-direction are all affronts to God.

The Question

Given this belief system, young male Muslims, especially those schooled in the madrassas, cannot help but view the current status of Islam in the world as humiliating. Their religion is meant to dominate and with it their culture. How can it be that so many Muslims live in societies filled with misery and squalor? Why is it that even the massive transfer of wealth from West to East for oil, surely a gift from Allah, has failed to lift so many Islamic countries out of their obvious morass?

With immense wealth could have come power to create an Islamic Renaissance that would dazzle the world, as the culture of Islam had once outrivaled the culture of Medieval Europe. Oil money might have built educational institutions capable of schooling male children (at least) in math and science, no matter how poor. It might have gone to fight poverty and squalor. It could have prevented famine. It could have supported multitudes of fledgling businesses, entrepreneurial enterprises that would grow to rival those in the West and provide beneficial products to the world. It could have gone into a blossoming of the sciences and the arts throughout the Islamic world. Why, by and large, has that not been the case? The question has naturally led to a sense that something has gone terribly wrong and that someone has to answer for it.

Some blame oil itself, calling it the devil's excrement, because its despotic possessors are often so thoroughly corrupted by it. Certainly the dictators who possess oil have had a much easier time ruling since they are not dependent on their people for revenue and can afford to equip large armies and police to repress and control their population. Nor do they usually have to worry about outside interference so long as they contain the repression within their own borders. On the contrary, they can count on buying the tools of repression in return for oil. Such is the power of money. That is not, however, a very satisfying answer to many people in the Middle East. More acceptable to many, especially those schooled in extreme Islam, is the idea that Western values and actions are to blame.

Western Imperialism and the Crusades

Not surprisingly, the Islamic dictators currently in power have had tremendous incentives to divert attention from their own misrule. They have urged their subjects to look outward for the cause of their misery to those who in Allah's eyes should be their subordinates.

They have had a very receptive audience, especially among extreme Islamists, but their arguments have become a sword that cuts two ways. The extreme Islamic fundamentalists have branded these tyrants Western lackeys because they have not taken oil money as a triumphant sultan or caliph would take tribute from a conquered and submissive foe, but as subordinates in need of Western know how and protection. In the view of extreme fundamentalist Islam, this humiliation alone has made these despots worse than infidels. As the claim that they are Western lackeys has gained currency, the more Western leaning, oil rich Islamic tyrants have sought even greater Western protection from the extremists they fostered and protected, even as they have continued to denounce the West and Western influence as the real cause of decline within Muslim societies. In common with their opponents, they rest their argument on a plausible interpretation of history, just as Marxists had done in their campaign against Capitalism. The Middle East, they point out, has been occupied and exploited by Western countries since Antiquity. Ptolemy, a Greek general in the army of Alexander the Great, seized the throne of Egypt and his descendents ruled as Pharaohs until Cleopatra VII, the last Ptolemy and the last Pharaoh of Egypt, was forced from the throne by Rome.[10] Imperial Rome controlled much of the Middle East by the time Christ was born, and Christianity was imposed as the state religion on the Roman Empire including the Middle East by the emperor Constantine.

The Crusaders descended on the Middle East during the Middle Ages, determined to recapture Jerusalem from the conquering Muslims. The French under Napoleon occupied Egypt; England and France carved up much of the Middle East after World War I, creating colonial possessions they often ruthlessly exploited.[11] The West made the Middle East a battleground during World War II and subjected the peoples of the Middle East to puppet dictators. The West continues to meddle in the Middle East which stokes a feeling of humiliation. And because of its towering need for oil, the West threatens the Middle East with future war and the desecration of sacred lands.

This history, seen as the basis for the misery and decline in Islamic societies, filters out the actions of favored local tyrants and religious fanatics; it also ignores significant portions of the historical record that reveal Islamic countries to be as aggressive as the Western ones they condemn. The missing history is not forgotten but presented almost as an alternate reality, reconstructed as a glorious past.

The Golden Age of Islam

During the seventh century A.D., around one hundred and fifty years after the fall of Rome (476 A.D.),[12] the Muslim followers of Mohammed began their conquest of the Middle East. At that time the Middle East was still under the influence of Christianity, the state religion of Constantinople (Byzantium). Muslims subdued the Middle Eastern Christians and forced those that would not convert, into dhimmitude or into "the next world." Under Ottoman Caliphs, Muslims spread into the Balkans subduing the Christian Orthodox and Catholic peoples, the Bulgarians (1371) and the Serbs (1389). They pushed farther into what became Austria-Hungary. At the same time, the Ottoman Turks subdued Muslim Arab lands, with equally brutal efficiency.

Alarmed, Europeans, still recovering from the Dark Ages that came after the fall of Rome, mounted crusades against the Ottomans in 1366, 1396, and 1444, but failed. The Ottomans continued their conquests. They conquered Byzantium, the last vestige of the old Roman Empire in the East and the seat of Christian Orthodoxy, in 1453 A.D. They overran Greece in the 15th century, taking Athens in 1468, and took Armenia in the 16th century, both Christian Orthodox countries.

Nor were the Turks the only Islamic people to penetrate into Europe. For nearly seven centuries, until the time of Columbus, Muslims from North Africa occupied a large part of Spain (712-1492 A.D.). And Muslims under Abd-ar-Rahman pushed into France with the aim of conquering the rest of Europe.

Retreat, Humiliation, and Redemption

Abd-ar-Rahman's army was confronted near Tours, France where it was stopped around 732 A.D. In Eastern Europe the sway of Islam gradually waned as the Ottoman Empire lost its grip. But even as the Ottomans withdrew from the Balkans, they left converts. The last to revert to Western control, Greece and Armenia, did not regain independence from the Turks again until the 19th and early 20th centuries respectively. Both suffered a blood bath. Byzantium remains part of Turkey. Constantinople is now Istanbul. Its famous Christian church Hagia Sophia was converted into a mosque in 1453. It is now a museum, though extremists want it to be used again as a mosque, a provocative symbol of Muslim dominance.

These were wars of aggression and conquest just as imperialistic as any devised in the West. But in extreme Islamic fundamentalist doublethink, Islamic aggression against European Christianity during Europe's weak and dispirited Dark and Middle Ages is not to be regretted or condemned but revived and emulated because it was divinely inspired. In contrast, Western Imperialism is to be condemned not because conquest is bad but because Christian conquests, unlike those of Islam, oppose the Will of God.

Since the retreat of the Ottoman Empire and the removal of the last Caliph from Turkey in the 1920s, Muslim countries have endured increasing, humiliating retreat that has led to their present state, according to the extreme Islamic fundamentalists. But these observations don't explain why this retreat occurred, and why the West has grown so dominant contrary to "God's" plans.

To explain that, the extreme Islamic fundamentalists provide an answer with roots in their ancient text and its interpretation of cause and effect: the defeat and humiliation Islamic societies have suffered are the result of their straying from the will and word of God; the state of Islamic societies across the world has come about because Muslims have allowed Western values, including democratic ideals, to infiltrate their societies. This permissiveness has led to a diminution of the true faith and traditional observances. Triumph over the infidel can only come through the renewal of the True Faith, which will grant to true believers knowledge of their enemies' weaknesses and protection from their enemies' strengths. True Faith will transform oil from the devil's excrement into the great gift Allah had intended to bestow upon them in their war against the infidel.

Time Past and Time Present

Memories go back far in the Middle East. Time past and time present intermingle. Extreme Islamic fundamentalists and anti-democratic dictators mix loathing of Western conquests with desire for Islamic conquests in the name of Allah and for the sake of power, power that could change the world.

Like Bin Laden, Saddam Hussein typifies this fever even though he was perceived as a secularist. A tip off was his representation of himself as a modern day Saladin. Saddam emphasized the connection between himself and Saladin on an Iraqi stamp with side-by-side portraits of himself and Saladin. It is Saddam-Saladin the conqueror and unifier of the Arab world who will defeat the West.[13] Why connect himself to Saladin? Saladin, who was born in Saddam's hometown Tikrit, not only succeeded in defeating Christian crusaders, he also conquered Jerusalem and temporarily united Egypt, Syria, and part of what is now Iraq, under one rule. It is an ancient dream which nurtures a modern temptation. Defeat Shia Iran, annex Kuwait, subdue Saudi Arabia, and what is gained? A hand around the jugular of the world; control of almost two-thirds of the world's easily accessible oil supply. The lure of such power can increase both megalomania and paranoia, the consequence of aspiring to possess what so many covet. Saddam was attacked. His aspirations were not fulfilled. He was not worthy, the extremist Islamic fundamentalists would say.

The Flash Point

Giving a focal point to the carefully stoked humiliation and anger in the Middle East is the state of Israel. As Muslim power and ambitions have grown, thanks to oil, waves of anger have churned increasingly out of control, widening in scope and intensity, while the conflict between Israel and the Palestinians has grown ever more dangerous.

The very fact that Israel exists on what was once Palestinian land has been a great source of anger and humiliation in the Islamic world for decades. The anger, resentment, mistreatment, murder, and vilifications of individuals on both sides of the Israeli-Palestinian conflict have not improved the situation. Yet mistreatment and gross violations of human rights are hardly unusual within many Islamic societies. The conflict remains unresolved, not so much because of the perceived theft of land but because a non-subjugated, non-Islamic people exist near the spiritual heart of the Islam world. That is what is truly intolerable to the extremists who would perpetuate conflict. It is also a godsend for anyone who would stoke the anger, resentment, hatred, and paranoia that make demagoguery possible. Neither the extreme Islamic fundamentalists nor the Arabic despots who control Arabia have had much reason until recently to end a conflict that serves them so well, so long as Israel exists.

What is seldom discussed is that the hapless Palestinians have never received sufficient money from the Arabic states to turn their refugee camps and ghettos into prosperous communities, though money has been routinely sent to families of suicide bombers.

Instead, Western countries have given the Palestinians most of the money they subsist on. Unfortunately, too much of that money has been squandered or misappropriated by Palestinian authorities. Before Arafat died, he allegedly funneled millions of dollars provided by Western

OIL AND THE THIRD WAVE OF TOTALITARIANISM   49

countries into private bank accounts. His successors are likely to follow the same pattern if they get Western money and the providers of the money fail to regulate its use.

Robbed of prosperity, the Palestinian people have been subjected to a bleak vision of the world that can not help but motivate them to perpetuate Jihad against Israel. It is a way of thinking that has surfaced many times in history, in medieval Europe and Japan, and in many other places where dire poverty, misery, and resentment have prevailed among the masses. It is characteristically otherworldly, death oriented, and escapist. Earthly life is short, miserable, and worthless, the instructor observes. How could it be otherwise today, which is so far from the Golden Age? What is our recourse but escape or triumph? Death, triumph, or some sort of apocalypse is the passage to a better life for the chosen. To merit the better life, the chosen must follow the wisdom of the few or of the one who is knowledgeable. The supreme leader possesses an understanding that surpasses that of the masses because it comes from God. What the leader offers to his followers is the restoration of God's law (as the leader interprets it) in place of the laws of man.

The Embargo

The oil embargo of 1973 was a physical manifestation of the resentment and anger, poverty and despair that continue boiling over. It was also a turning point in Middle Eastern perceptions of the West, which had appeared indomitable since the nineteenth century.

The initial impetus for the embargo was to mute the Western response to the Egyptians and Syrians as they amassed on Israel's border with the intent of invading Israel and exterminating it. The possibility of a massacre was great, but was never realized. The Egyptian and Syrian armies failed in their mission. After it became clear that the attacking Arab armies could not "wipe Israel off the map" as planned, the Egyptians and Syrians demanded that Israel withdraw from lands it had occupied in the 1967 war. The Arabic members of OPEC reinforced this demand with their embargo.

The thinking behind the demand for Israel to withdraw is not hard to tease out. If the Arab countries could not get all of what they wanted (the end of Israel) they would demand part, which they expected Western countries to agree to as a gesture of appeasement. Why should they expect the demand to be met? There are several reasons. Although they had sought to destroy Israel, they now pretended they only wanted the lands back that Israel had won the last time it had been attacked. The demand was not so extreme that it would be readily refused, given it was made not of Israel but of the Europeans and Americans who were not directly involved in the conflict but could influence the outcome. The "reasonableness" of the demand was something Westerners could accept so that they would not have to face the naked fact that OPEC was threatening to cut oil supplies if they didn't submit to OPEC's demand. Its language gave the West, as well as the humiliated Arabs, a way to save face.

The embargo demonstrates the power of those who control oil over those desperate to consume it. Noticeably absent was fear of military action by the West. In October 1973, Libya, Saudi Arabia, and its oil-rich neighbors voted to cut off supplies to America. In this policy, intended to divide Europe from America and to put pressure on the United States, the Saudis sought to reward compliant Western countries that sided with the Arab states.[14] They demanded that American oil companies ship oil to favored Western European states while denying it to the United States. Under threat, the Western Europeans hardly seemed to notice that it was the Arabic states that had been the aggressors. The U.S. received little overt assistance or even verbal support from most of its European allies in its efforts to save Israel. A notable exception was the Netherlands.

The European press, by and large, expressed the opinion that the U.S. was siding with Israel against European interests. Whether they would have changed their mind had a massacre of Israeli Jews actually occurred can never be answered with certainty.[15] The Europeans had a real reason to fear that an embargo could result in economic ruin. At the time, the U.S. imported 12-14% of its oil from the Middle East and could still produced enough oil to mitigate loses, while Europe imported something like 80% of the oil it used. Was Europe willing to invade Saudi Arabia and the other oil producing countries to get oil? Could they prevent sabotage of the oil fields if they did? Had they suffered an embargo like that imposed on the Americans, the result would have been far more devastating than it was in the U.S.

As it turned out, the oil shock that followed the embargo, because of its suddenness, caused the retail price of a gallon of gasoline in the U.S. to rise quickly from a national average of 38.5 cents in May of 1973 to 55 cents in June 1974. Shares of stock that were traded on the New York

Stock Exchange, excluding oil stocks, lost $97 billion in value in just six weeks. By the summer of 1974, oil imports were down 7% and Americans were queuing up in long lines for gasoline. The decrease in imports could have had a more devastating impact had oil not been sold on the open market to America. In Europe the price rise that accompanied the oil embargo against America was at least as bad as the actual embargo was in the U.S.

In November 1973 the members of the European Economic Community (EEC) issued a statement calling for Israel to withdraw from the land it had taken during the 1967 war, as the Arabic OPEC members had demanded. OPEC duly lifted its embargo against target members of the EEC, principally the Netherlands, and reduced the price of oil. Separately, the Europeans also began a dialogue with the Arab states that led to an increase in legal immigration from North Africa and has since raised hopes among extreme Islamic fundamentalists of a re-conquest of Europe from within.[16]

In March of 1974 after negotiations in Washington, the embargo against the U.S. was lifted despite the U.S. refusal to abandon its support for Israel or agree to the other terms. Fortunately, OPEC had not been able to sustain its action without suffering its own economic distress. But there were political not just economic reasons for them to relent. The Saudis and other dictators in the region were becoming increasingly worried about being overthrown from within or attacked by other states in the region. They needed Western protection just as the West needed Middle Eastern oil. Still the effects of the energy crisis lingered in the West throughout the rest of the 1970s.

Initially, the oil shocks of the 1970s motivated us to action. We beefed up our domestic supplies, diversified the sources of imported supply even more, embarked upon a massive conservation effort that included buying smaller Japanese made cars, and began to create an alternative energy industry. These measures, if continued and expanded, could have effectively prevented the Persian Gulf oil states from using oil as a weapon in the future – but most of these measures were not sustained. A significant dip in oil prices between 1985 and 1999 quashed many of the alternative energy companies that had sprung up in the U.S. and were making a difference. Memories faded, conservation measures diminished, and the warning the embargo gave us ultimately was forgotten. Economics, not common sense, ruled.

Today an embargo could produce a very different result. Hopes of averting the effects of another embargo are now pinned largely upon diversifying oil supplies. America looks to Canada and the Americas for help. Europe is forced to consider Russia, which is breeding megalomania and paranoia in Moscow. Plausible fears of conflict over dwindling resources are already being used in Russia to stoke fear of external enemies and to brand any dissenters as agents of foreign governments that covet Russia's oil. With the aid of oil, Russia is reverting once again to its past, evoking its dismal history of repression and totalitarian rule.

Even if successful, attempts to escape the influence of Middle Eastern oil by diversification are not likely to do much good in a real crunch. Today a significant cutoff of supplies in the Middle East will have a strong ripple effect throughout the world's unprepared economies, much more so than it did in 1973. And the threat is likely to grow as reliance on easily accessible oil outside the Middle East reduces those supplies, effectively concentrating more of the world's easily accessible oil reserves in the Middle East.

So how likely is another embargo? It's hard to say. As it stands today, the oil consuming nations remain dependent mostly on the less extreme of the Middle Eastern despots, but that could change.

The Utopian Disguise

Were the extreme Islamic fundamentalists of Al Qaeda to obtain their aims in Iraq, in Saudi Arabia, Libya, and elsewhere, they would control enough of the world's oil supply to make the situation in the West critical. However, the people who are experiencing the initial effects of extreme Islamic fundamentalist ambitions are the Muslim populations themselves.

The consequences of an Al Qaeda victory are predictable. Dissent by moderate Muslims and non-Muslims would be completely snuffed out in the Middle East. So would the moderate argument that the Koran does not condone, much less demand, the indiscriminant murder of women and children, the torture and beheading of innocent kidnap victims, or suicide as a means of murder. The extreme Islamic fundamentalists have already shown they will not hesitate to use the same terror tactics against dissenting Muslims as they use against the West.

Once established, Bin Laden's utopia, led by a supreme leader, would rain death down upon its citizenry through fatwas aimed at individual subjects. These fatwas would make murder not simply legal but an obligation of all Muslims who are in a position to carry it out.[17] Death would be demanded of targeted individuals for any number of offenses starting with the obvious: death for offending the prophet or the Koran, death for being an unsubjugated infidel, death for seeking to convert a Muslim to another faith, and death for ceasing to be a proper Muslim. Standards for determining such offenses can easily be set low. Imprisonment and torture of anyone perceived to be a threat to the regime would increase dramatically. Mass murder in the Sudan, mass slaughter in Algeria, exterminations in the horn of Africa, are all forewarnings of this outcome, which are apparently going unheeded in many Muslim communities.

However, it is not the present that foretells this future so much as the past. Unfortunately, we have seen all of this before. Different as it may seem, the Utopian vision of Bin Laden is but a variation of the vision of other totalitarians bent on controlling every aspect of other people's lives. The terror perpetrated by those in control only begins with the identification of the "enemies of the people." Whether they be "class enemies" or "conniving racial inferiors" or "infidels," it hardly matters. Building on paranoia and uncertainty, often enhanced by whimsical rules of obedience, the totalitarian leader turns friend against friend, men against women, children against their parents, until nearly everyone is so fearful for his or her life that collectively they fear to think an original thought much less speak out in public. What is most sad about this type of government is that many people, given little other choice, are likely to buy into the whole psychology that represses them. They will expose others as traitors to prove their own fidelity. They will glorify the regime, its thugs and sociopaths, in the hope that such behavior will keep them safe, while the thugs and sociopaths revel in the intoxicating power they would never possess in a saner society.

Unfortunately, humans are not very good at learning from the past. If we were, we would not once again have to deal with ideas that can transform our world into a slaughterhouse where genocide, suicide, and terror prevail and a dark inquisitional fervor shrouds the light of reason.

The extreme Islamic fundamentalists of the third totalitarian wave cannot be appeased, any more than the fanatical Nazis and Stalinist Communists could be, so long as their ideology blinds them. Neither humiliation nor success moderate fanatical ideologues. The extreme Islamic fundamentalists will interpret defeats as attempts to humiliate not them but Islam. And they will see their successes as evidence that history and God are on their side.

Perception is everything. Reality often counts for little in the Middle East. Rather than recognize the dangers of extreme Islamic fundamentalism, the young and politically naive who join the Jihadist movement are likely to have bought the argument that Western values and imperialism are to blame for Islamic impoverishment and failure and that they pose an imminent threat. If both physically abused and taught to be violent as children, their conviction may be reinforced through a projection of blame for their treatment onto the outside world. Once lured into Jihad, would-be Jihadists are likely to stay engaged even if they recognize its extreme inhumanity and begin to doubt. To break away would be dangerous. To stay holds the lure of personal power that comes with the fear their violence breeds. Everything they do can be justified as necessary for conquest by the Faith and furtherance of God's will. And God's eventual triumph can be convincingly argued as inevitable, given the West's dependency on foreign oil, its decadence, its apparent indecisiveness, and the existence of its own indigenous strains of Utopian totalitarianism.

Nonetheless, inevitably mass murder and terror will reveal themselves for what they are to all but the most ideologically blind, not the means to Utopia, religious or otherwise, but mechanisms of control, levers of manipulation for those drunk on power who would change the world then punish it for not being what they hoped it would be.

The Response of the West

The Western response during the Embargo exposed Western weakness and divisions. Since September 11, 2001, the divisions, fears, and threats have only grown, exacerbated by oil money. Consider how much Iran's nuclear defiance is helped by oil or how much the danger of the North Korean nuclear program has been magnified by terrorists with oil money to spend on nuclear weapons. Consider the likely consequences of the Taliban recapturing power in Afghanistan or prevailing in Pakistan thanks to money from Bin Laden and certain Saudi oil princes. Al Qaeda's leaders would have easy access to an extensive training ground for their foot soldiers, to biological weapons in Russia's breakaway republics, to Pakistan's nuclear weapons and its nuclear experts, to the lucrative opium trade, and to laboratories where they might tinker with their weapons of mass destruction. Then consider the attempts to counter Islamic extremism by imposing democracy on a region that views Western democracy with religious antagonism. It recalls the attempts to create democracy in Germany after World War I when democracy was associated with the conquerors. In what could become a multi-front, global war with the oil region of the Middle East at its center, Bush's Wilsonian counterthrust remains an enormous gamble with an uncertain payoff.

The jury of history convened to determine the wisdom of this approach may still be out. But one thing is already apparent; we are fighting the type of urban war that Bin Laden had hoped for. We are fighting a war on his terms, based on what he learned from the Communists and Nazis about the best strategy and tactics to use in fighting democracy. In his quest to defeat the West, he and his associates have improved upon the Communist cell and Nazi fifth column, as a supreme weapon of urban warfare. Waging urban warfare through cells that are stateless and loosely connected neutralizes the advantages of big armies and advanced technology. It allows Al Qaeda to attack our fragile energy and economic infrastructure at a fraction of the cost required to defend it. The use of urban warfare is even more effective when the cell exists within a compliant community. By retreating into such a community after attacks, his extremists can count upon a response from their enemies that will antagonize the local population among whom they hide.

The Tar Baby Strategy

Bin Laden was also emboldened to draw the U.S. into this war of attrition, as he characterizes it, after observing the repeated behavior of the U.S. in Vietnam, Lebanon, Somalia, and elsewhere. He was further emboldened by what happened to the Soviet Union in Afghanistan and afterwards. He believes the breakup of the Soviet Union was due to its defeat in Afghanistan, which he attributes to himself and his followers and to God.

Bin Laden's evolving tactics now rely upon keeping the U.S. engaged in a costly Vietnam type of war, in which his soldiers can retreat into friendly safe havens after ambushes and in which his soldiers become indistinguishable from those among whom they live.

What is most likely to keep us engaged in his war of attrition more than the desire to destroy Bin Laden himself is Middle Eastern oil. If Iraq is further destabilized, the chaos could spread. Oil production across the region could be jeopardized. Massive disruptions in the importation of oil to consuming nations could throw the world economy into chaos.

Our increasing involvement in the Middle East since the Iraqi invasion has given Al Qaeda an unparalleled opportunity, one that recalls a nineteenth century American folktale about a clever fox and a self-assured rabbit. In the tale the fox figures out a way to take down the rabbit by playing on his weaknesses. Chillingly, the fox creates a tar baby – a statue made of oil tar – and places the figure right smack in the road that the predictable rabbit is likely to take. The rabbit once he encounters the strange figure is unable to resist becoming stuck in it.

The biggest uncertainty in Al Qaeda's tar baby strategy so far is not what the U.S. will do but what the Iraqis and Iranians will do. Bin Laden's group would certainly like to see the U.S. attack Iran too. To Sunni Islamic fundamentalists, Iran is in a state of apostasy, which is worse than being an infidel. But the extreme Sunni Islamic fundamentalists can hardly count on instigating a conflict between the U.S. and Iran, which is one of many reasons why Al Qaeda may have chosen instead to encourage the brutal conflict between Sunnis and Shia[18] in Iraq. They have long hoped to "awaken the Sunnis" from their "slumber" and radicalize them. But Al Qaeda's strategy may have unintended consequences. Some Sunnis have already turned against Al Qaeda, while many of the Iraqi Shia have strengthened their ties with Iran. If the Shia majority in Iraq win the civil war, Shias could extend their regional influence, already formidable thanks to oil. The consequence could be a widening of the great schism in Islam and violence in the Middle East on a scale not yet seen with an outcome that favors the Shia. Given the shifting political sands of the Middle East, other stranger outcomes are also possible. Despite Iran's Shia apostasy – from Al Qaeda's point of view – the Sunni and Shia extremists could make a temporary alliance in order to intimidate more moderate Muslims.

No poll has been taken, but almost certainly the majority of people across the Middle East want peace. Would-be dictators and extreme ideologues, in contrast, seek violence and chaos. Exploiting social divisions and instigating violence is how they often gain power. Divisiveness and disorder radicalize and polarize a population. With their security threatened and passions inflamed, people of any country can come to believe that the instigators of violence are their defenders. Under this illusion, they may accept the "defenders'" radical solutions, including repression, if it means the defeat of their perceived or actual enemies and the establishment of some sort of order and security.

Defeating the Tar Baby Strategy

To counteract this psychology, the people who are being violated must first realize who the real instigators of the violence are and be ready to turn the tables on them while it is still possible. In Iraq, three goals must be achieved to avoid further civil war and bring peace without dictatorial domination: 1) foreign Jihadi fighters must be shut out of Iraq, which requires some cooperation from Iraq's neighbors, 2) the Sunnis must be persuaded to fight Al Qaeda rather than the Shia, and 3) the Shia government must be persuaded to crack down hard on any resurgence of the Shia militias, especially the so-called Madhi Army of al-Sadr, a man whose dark passions are readily discernable. It is crucial that the instigators of violence be combated by their perceived constituents, the people they appeal to for support. Shias have to stop Shias' violence. Sunnis have to stop Sunnis. Otherwise, the violence will be seen as partisan and is unlikely to be overcome. Defeat the true instigators of violence on both sides and the violence will stop.

As outsiders, we could help mediate between the Shia and Sunni in order to coordinate their response to their own extremists. We could argue that without democracy the alternatives for Iraq and the Middle East are grim: corrupt, repressive dictatorships or extreme Islamic fundamentalist states dreamt of by people who think nothing of blowing up innocent civilians going about their daily routines or of cutting the heads off helpless captives.[19] But will the people listen and will they unite to fight those who would pull them apart? And if they don't, where does that leave us?

We cannot choose for the peoples of the Middle East what they will not choose for themselves. We cannot decide for them what they will become. But their decisions can affect us. So wishful thinking aside, given the volatility of the Middle East, we should make the Middle East as irrelevant to us as possible. We can best do that by drastically reducing our dependency on imported oil.

Our oil dependency magnifies the power and influence of extreme Islamic fundamentalists just as it saps and polarizes the West. Oil money fires the ambitions of despots, increases aggression, and enables greater repression. Oil dependency makes it possible for oil producing nations to blackmail oil consuming nations, to manipulate the foreign policy of consumer nations, and to pit consumer nations against one another. It limits our options, makes our actions more predictable, encourages corruption and bloodshed,[20] and sparks conflict.

Moving to end our oil dependency will not end the threat from Islamic extremists and that of other totalitarians. But it will make a tremendous difference because it will take away the economic threat of oil dependency at the same time it goes after the money supply that, like blood nourishing a cancer, helps tyranny to grow. And it is an action that is entirely internal to our country, requiring no army and no bloodshed. Unfortunately, it can not be done over night, which is what makes our predicament so difficult to correct. But it will start to have an effect immediately if we act to make oil import reduction a high priority and if we recognize that a tar baby strategy is being used against us. We have expended far too much of our resources, human and monetary, on the Middle East since 9/11, while leaving our oil dependency unchecked. We dare not let this situation continue. Reducing oil imports will defeat the tar baby strategy, while maintaining or increasing our oil dependency is likely to mire us even more deeply in the swirling uncertainties of the Middle East.

6

The Environment, the Mad Hatter, and His Dirty Dishes

The hydrocarbon fossil fuels – crude oil, natural gas, and coal – have rightly been called the blood of modern civilization. An amazingly wide variety of products, mostly power related, are extracted from them. We encounter many of these products on a frequent basis: propane, methane, ethane, butane, naphtha, gasoline, kerosene, jet fuel, diesel oil, lubricating engine oils, heating oils, fuel oil, and a variety of residuals that derive from oil such as coke, asphalt, tar, and waxes (used in plastic, crayons, and other products) sulfur, mercury, lead, and hydrogen gas.

Unfortunately, as the political dangers associated with oil have increased, so have the environmental dangers associated with hydrocarbons. Oil and coal are particularly dirty, especially when combusted. There is clear evidence that they have already degraded the quality of life on Earth. And there is a reasonable expectation that the problem will get worse, despite important measures already taken by the industrialized nations to reduce pollution.

The Hydrocarbon Life-Cycle

Pollution from oil and coal occurs throughout the hydrocarbon life-cycle we've created, starting with the extraction of crude oil and coal from the ground. In the case of crude oil, seepage and spills at the wellhead or from under the ocean floor are infrequent but dramatic and expansive. As a small child, I recall getting tar on my feet and smelling tar whenever I walked along Hope Ranch beach near Santa Barbara; so much so that at age three I thought the ocean was supposed to smell like tar.

Later I watched as that association of tar with the ocean became more grotesque. On January 29th, 1969, six miles off the coast of

Summerland, California, Union Oil Company workers were busy drilling a well 3500 feet below the ocean floor when a blow out occurred. The oilmen immediately attempted to cap the hole to stop the venting of natural gas, the cause of the blow out, but their attempts failed. Cracks opened and expanded along the ocean floor following a fault line near the well. Oil oozed from the cracks in increasing amounts together with natural gas. The oil workers struggled for 11 days to stop the rupture as over 200,000 gallons of crude oil bubbled to the surface and spread, forming a huge oil slick. Beaches from Rincon Point to Goleta some thirty miles up the coast were inundated as were the Eastern facing beaches of the Santa Barbara Channel islands: Anacapa, Santa Cruz, Santa Rosa, and San Miguel.

Looking seaward from Rincon Point, I could see the ocean turned black with oil far into the distance. Its sluggish waves seemed unearthly, like the waves of a petrochemical sea on another planet. With each incoming tide, the corpses of seals and birds rolled ashore. The dissection of the seals revealed their lungs were clogged with oil, causing respiratory failure. Birds like plovers, godwits, and willets that depend on sand crabs and sand fleas along the shore escaped, but birds that dove beneath the ocean waters for fish were covered with tar from beak to tail. Caustic detergents used to disperse the oil slick added to the misery of the wild life. The chemicals robbed birds' feathers and seals' fur of natural waterproofing and burned their skin.

An investigation later found that the Union Oil Company's platform had inadequate protective casing. Steel pipe sheeting inside the drill hole would have prevented the rupture, but the oil rig did not have to conform to state regulations that would have mandated the casing because the platform was beyond the three-mile coastal limit. Union Oil lost millions of dollars from the clean-up efforts and from payments to local fishermen and local businesses. Fred L. Hartley, the president of Union Oil Company, was quoted as saying, "I don't like to call it a disaster because there has been no loss of human life. I am amazed at the publicity for the loss of a few birds."[1]

The accident off Summerland was neither the first nor has it been the last of its kind around the world. What it did was help stimulate environmental activism. The next year Earth Day was born nationwide, though Earth deserves more than a day of attention, as environmentalists themselves would argue.

Spills like that at Santa Barbara have been unintentional as well as infrequent. More worrisome are the deliberate acts of sabotage that are as heedless of the environment as they are of human life. In 1991, over 600 Kuwaiti oil wells were deliberately set ablaze by Iraqi[2] soldiers, causing a smoke blanket that blocked the sun. Huge amounts of oil from the runaway wells pooled, then seeped into the ocean, combusted, permeated the air, filled the lungs, and eventually percolated down to the aquifers. It is an act that is likely to be repeated. Oil wells are simply too easy a target.

The transport of oil from fields to refineries, whether by pipeline, tanker, oil barge, or oil trucks, provides more opportunities for spills. The Exxon Valdez spill in Alaska is a prime example. Even worse, oil tankers docked at major ports and gas and oil trucks traveling the highways are bombs waiting to be exploited.

Pollution from coal extraction has never been as dramatic, but it has probably killed more people outright. Coal mining accidents are not frequent but dramatic. Often more deadly are the fine particulates in coal dust. Particulates of almost any kind that enter the lung adversely affect respiration. Coal dust particulates, which are laced with other substances such as arsenic, mercury, and sulfur, are particularly deadly. They cause a disease that has plagued coal miners for centuries. Known colloquially as "black lung," it is still a danger to miners in many countries.

Potentially as hazardous are some contaminants whose existence in coal deposits might come as a surprise. Most coal contains radioactive radium-226 as well as uranium, thorium, potassium-40, and lead-210. The total levels of these contaminants should be generally about the same as in other rocks of the Earth's crust. The problem for miners is concentration. For example, radium-226 decays to produce radon-222, which accumulates in mines. Normal environmental levels of radon[3] range from 4 to 20 Bq/m^3. In mines and caves, the concentration of radiation can be over 1000 times higher. The constant exposure of miners to such high levels of radon's alpha particles over time increases the chances of many serious respiratory problems and possibly cancer. Furthermore, radioactive material often piles up as waste in and around a mine and becomes concentrated. These toxic waste dumps are one of the many hidden costs of hydrocarbon use, which may be estimated in the billions if not trillions of dollars.

The pollution problems multiply when oil and coal are heated to high temperatures during the next stage of their hydrocarbon life-cycle. Coal is simply combusted to generate electricity at power plants, to provide heat for homes, and to fire furnaces with which iron, steel, and aluminum are produced.

Oil heating is more complicated. Oil is heated in refineries to convert it into useful products. To really understand the pollution problems caused by oil due to refinement, you need to understand the refinery processes. The type of crude oil dictates some of the processing requirements and the amount and types of wastes produced, so we'll start with oil typing.

Crude oil is typed or categorized mainly by its viscosity and color, which vary significantly according to type. The color ranges from almost clear to a dark tar black, while viscosity ranges from almost watery to almost solid. Generally, the darker the oil, the more viscous it is and the greater the number of impurities it has.

The Page Museum at the famous La Brea Tar Pits in Los Angeles is a good place to get a feel for really viscous oil. At the Museum, visitors are asked to lift a rod stuck in a sample of the gooey brown black oil from the pits. The effort required to lift the rod is impressive and evokes images of Pleistocene creatures struggling in vain to extricate themselves from an oily morass deceptively covered by the water they or their prey meant to drink. The image is aptly metaphoric.

Awareness of the basic characteristics of crude oil are important in making an initial judgment about its composition and therefore about its quality. The dark, thick, viscous appearance of the oil found in the La Brea Tar Pits betrays its composition, mostly of asphalt and tar heavily laced with impurities. It is very low quality oil, which is one of the reasons why it has been left as a tourist attraction.

Viscosity and color give important hints about the composition of the oil, but at a deeper level the various hydrocarbon chains that compose it determine what is contained in a particular oil sample. Hydrocarbon chains can be classified according to the number of carbon atoms they have. For example, petroleum gas, which is used for heating and making plastics, has 1-4 carbon atoms in its chain. Gasoline has 5-12 carbon atoms; diesel around 14 carbon atoms; and fuel oil, which along with coke is used for making steel, contains chains with 20-70 carbon atoms.

Understanding hydrocarbon chains is important to refiners for two reasons. First of all, as mentioned, different chains represent different substances that can be extracted from oil. Second, the chains have different boiling points and therefore different temperatures at which

THE ENVIRONMENT, THE MAD HATTER, AND HIS DIRTY DISHES   63

they evaporate and distill or condense back into liquids. As an example, the boiling point of gasoline is between 104°F and 392°F, while the boiling point of diesel is between 482°F and 572°F and that of kerosene is around 392°F. Refiners exploit temperature differences using fractional distillation; that is, they boil crude oil to about 1112°F in a distiller in order to vaporize all the hydrocarbons, and then capture the vapors as the vapors re-condense at different temperature levels within the distiller column. The process is illustrated in Figure 5.

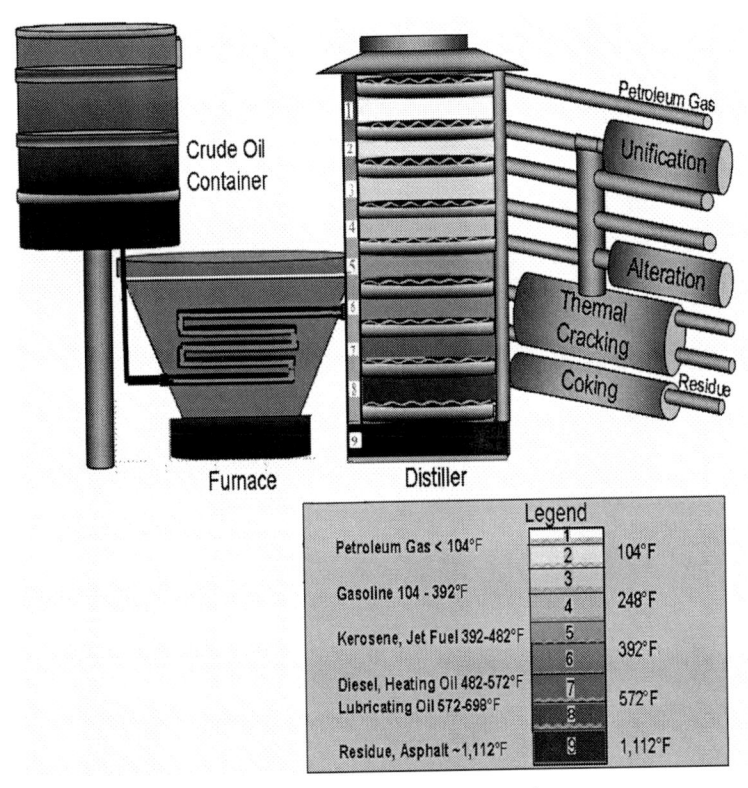

Figure 5. Oil Refinement[4]

The means by which the liquefied vapors are collected is through trays placed at the different condensation levels in the distiller column. Residuals such as coke, asphalt, and tar remain immune to vaporization at all but the highest temperature, 1112°F (600°C). They are the leftovers of the distillation process and are mostly condensed at the bottom of the column.

Fractional distillation, though primary, is only a first cut process. Condensing hydrocarbons are transferred by tube from the trays into separate condensers that cool them further. The hydrocarbon chains that are separated into "fractions" during condensation are next subjected to thermal cracking, unification, and alteration.

During these steps, the hydrocarbon chains are manipulated to increase the yield of the more valuable chemicals. The three processes used can literally transform one compound into another: 1) **thermal cracking** breaks large hydrocarbon chains into smaller ones, 2) **unification** combines small chains into larger ones, and 3) **alteration** rearranges pieces of a chain. All three processes use high temperatures (around 900°F) and catalysts[5] (such as hydrofluoric acid, sulfuric acid, or zeolite) which are chosen based on to the specific process needed to manipulate a hydrocarbon chain.

After distilling out and manipulating the hydrocarbon chains using cracking, unification, and alteration, the refining process purifies the fractions. Columns of sulfuric acid remove nitrogen compounds and tiny amounts of residual solids such as tars and asphalt, while nitrogen-sulfide scrubbers remove sulfur impurities. The final process is called blending, which combines various fractions to create the final products ranging from petroleum gas to gasoline to fuel oil.

The House of Toxic Horrors

Unfortunately, almost every chemical used in the oil refinement process or produced by it is harmful to life. Many are both mutagenic and carcinogenic, which means that they cause both birth defects and cancer. The list of chemicals used is a list of toxic horrors. Of these toxic chemicals, only a few are actually reported to the Environmental Protection Agency (EPA), which is charged with guarding our environment. The chemicals that are regularly reported are listed in the Toxics Release Inventory (TRI).[6]

The TRI report, which has been required only since 1986, excludes a number of chemicals. For example, 1,2,4-trimethylbenzene is reported in the TRI, while 1,3,5-trimethylbenzene is not. However, many of the chemicals excluded have structural and toxicological properties similar to those that are reported and are present in amounts predictable from the amounts of the corresponding reported chemicals. Other pollutants like carbon monoxide, nitrogen oxides, sulfur dioxide, and particulate matter, while not included in the TRI, are included in the EPA Air Quality System (AQS) database. A portion of the TRI list is shown in Table 1.

Table 1. Common Chemicals of Oil Refineries[7]

Crude Oil Components	Chemicals Formed During Refining	Catalysts or Catalyst Components	Other
Anthracene	Ammonia	Beryllium & beryllium compounds	Acetone (delisted 1995)
Benzene	Benzene	Chlorine	Diethanolamine
Cyclohexane	1,3-Butadiene	Chromium-chromium compounds	Glycol ethers
Ethylbenzene	Cumene	Cobalt & cobalt compounds	Hydrochloric acid aerosols
Phenol	Cyclohexane	Copper & copper compounds	Methyl ethyl ketone
PAHs (1995 listing)	Ethylbenzene	Hydrogen fluoride	Methyl isobutyl ketone
Toluene	Ethylene	Molybdenum trioxide	Methyl Tert-Butyl Ether
Xylene (all	Methanol	Nickel & nickel compounds	Phenol
Trace Metals:	Naphthalene	Phosphoric acid	Sodium hydroxide
Antimony	Propylene	Selenium & selenium compounds	Sulfuric acid aerosols
Arsenic	Toluene	Sulfuric acid aerosols	1,1,1-Trichloroethane
Chromium	1,2,4-Trimethylbenzene	Thallium compounds	1,1,2-Trichloroethane
Cobalt	Xylene (all isomers)	Vanadium (fume or dust)	Trichloroethylene
Copper Mercury, Lead Selenium, Nickel		Zinc (fume or dust) & zinc compounds	

Possibly the worst of the pollutants are the benzene compounds. Benzene is a first-rate carcinogen and mutagen. Benzene exposure has been linked to all types of leukemia. However, it is most associated with rare forms of leukemia including acute myelogenous leukemia (AML) and acute lymphocytic leukemia (ALL). Scientists have also linked benzene to non-Hodgkin's lymphoma (NHL), Hodgkin's disease, multiple myeloma, anemia, pancytopenia, cytopenias, myelodysplastic syndrome (MDS), aplastic myelofibrosis, and polycythemia vera.[8]

Exposure to benzene is particularly deadly when prolonged. Cancers can develop after only 5 years of exposure, but they can also take more than 30 years to appear. Long-term exposure may also adversely impact bone marrow and blood production. Benzene is not considered safe at any level.

It may not come as a surprise to learn that benzene's deleterious effects aren't a recent discovery. Benzene has a fairly long history in medical literature. As early as the 1920s, benzene was linked to leukemia.[9] The American Petroleum Institute noted in the 1940s that benzene caused leukemia and warned that any level of exposure to benzene posed risks. In 1977 a major epidemiological study of workers exposed to benzene further demonstrated the risk of leukemia. Yet little has been done because the effects of benzene like those of many other pollutants are insidious – not immediate – so no sense of crisis exists. Without a crisis, both the public and government are often slow to act.

The United States and some other countries, most notably Japan and the Western European countries, are attempting to reduce refinery leakage and to penalize the dumping of poisonous wastes from refineries into local rivers and oceans. In most of the rest of the world, little attempt is made to stop the dumping or reduce the leakage. Communist countries like China, and former Communist countries in Eastern Europe, are among the worst offenders. With little to keep them in check, they have polluted more with less. The worldwide failure to reduce pollution has meant that we've taken out a hidden mortgage. Eventually this hidden mortgage will come due and the ensuing problems will be paid for, not by the polluters, but by the public.

Each day, an average-sized refinery[10] produces more than 10,000 gallons of waste products. Each day, our air, land, and water are assaulted by over 100 pollutants emitted from the distiller stacks and from leaking equipment at refineries. And this pollution is only the daily pollution lost through inefficiency, a small percentage of the total pollution that ends up in the environment. The products of the refinery process that remain, products chemically altered to an extent that many are unknown to Nature, are now ready for consumption.

Unlike oil, coal does not require refinement to use, which has made it particularly attractive as a fuel for generating electricity and for running furnaces used in the manufacturing of many products. Because coal can be used right out of the ground and because it is abundant in the U.S., it seems cheap, but combustion of coal will not be cheap in the long run.

What should make a sane person hesitate to combust coal is what it produces. Coal combustion liberates many of the same pollutants found in crude oil: oxides of carbon and nitrogen, arsenic, mercury, sulfur, and methane. But it also adds others. It produces far more particulates than oil such as coal ash and fly ash, which can enter lungs and lodge there. The particulates in coal ash reflect the composition of coal. The particulates of fly ash are different. They are composed chiefly of non-combustible silicon compounds that form into tiny glass spheres. The tiny glass spheres that make up the bulk of fly ash attract heavy metals such as uranium. The uranium released to the atmosphere with escaping fly ash is about 1% of the amount of uranium that was in the coal, according to the National Council on Radiation Protection and Measurements (NCRP).[11] It would be much more except for the fact that since the 1960s substantial amounts of fly ash have been contained through the use of particulate precipitators, at least in U.S., Japanese, and Western European coal-fire power plants.

Coal ash is less well regulated. The coal ash and fly ash that don't escape into the atmosphere usually end up in huge cinder piles or in ash ponds at coal plant sites. This hazard at coal plants is often worse than the hazard inside and around coal mines because the radioactive material as a residue of combustion is even more concentrated.

The Rising Tide of Pollution

Though coal pollution is bad enough now, it is expected to rise significantly during the next few decades in the U.S. and even more dramatically in developing countries where the need for regulation of pollutants is more likely to continue being ignored. Alex Gabbard of Oak Ridge National Lab (ORNL) has graphed the rise in coal pollution since 1937 and projected the increase to the year 2040.[12] He believes that by 2040, over 637,409 million tons of coal will have been combusted. The increase in coal combustion is projected to cause global releases of radioactivity of 2.7 million curies.[13]

The Insidious Burden

While the projected increase in radioactive pollutants is worrisome, of greater concern is the accelerating trend in the use of all hydrocarbons. As hydrocarbon pollution accumulates and spreads, driven by the growth of human populations, it will impose an increasingly insidious burden on the Earth's ecological system.

Photochemical Smog

The use of oil products, the next phase of the oil hydrocarbon life-cycle, adds to the deleterious effects. The most visible pollution for many people is smog. Much of this pollution is due to the inefficiency of the gasoline fueled internal combustion engine.[14] However, smog also comes from refineries and factories that use hydrocarbons like coal, and vent the unburnt waste products into the atmosphere.

The main contaminants that produce smog are carbon monoxide (CO), nitrogen oxides (NOx) – which include nitrogen oxide (NO) and nitrogen dioxide (NO_2) – unburned gasoline, and volatile organic compounds (VOC), which are mostly hydrocarbons (HC) derived from unburnt fuel vapors. Carbon monoxide, hydrocarbons, and oxides of nitrogen react with sunlight to create ground level ozone, which is the major component of photochemical smog. This ground level ozone is a very unstable, highly reactive element that makes an ideal bleaching agent but not an agreeable inhalant. Its corrosive properties damage the bronchioles and alveoli sacs in the lungs. Especially susceptible are those with existing respiratory conditions such as asthma, the very young, and the elderly. In addition, carbon monoxide interferes with the blood's ability to carry oxygen to the brain, heart, and other tissues; it is particularly destructive because the blood actually absorbs CO more readily than oxygen. Those most at risk are children, the old, and people with heart disease.

Besides the main contaminants, secondary contaminants add bulk to the composition of smog. The best known secondary contaminants are carbon dioxide, sulfur oxides, and particulate matter. All are products of combustion.

Carbon dioxide (CO_2) is the leading greenhouse gas.

THE ENVIRONMENT, THE MAD HATTER, AND HIS DIRTY DISHES   69

Sulfur oxides (SO_x), another greenhouse gas, constricts air passages, making it difficult to breathe, especially for people with asthma and for children. Even brief exposure to relatively low levels of sulfur dioxide can cause an asthma attack in the most susceptible.

Particulate matter (PM) is a hydrocarbon product of unburnt fuel. Larger particles can be stopped in the nose and upper lungs by the body's natural defenses. The smallest particles escape the body's defenses and go deep into the lungs, where they become trapped.

Before the EPA clean air acts began to make a difference, newscasters in various cities adopted a color code from purple (the worst) to green to indicate smog levels. In the 1970s in major cities like Los Angeles, Houston, and Detroit, the number of summer days in which the smog levels reached purple levels increased alarmingly. Los Angeles, which is a natural catch basin for smog, was particularly affected. It is a testimony to what the government can do through legislation that this situation has improved rather than gotten worse, despite the increase in population.[15]

Elsewhere in the world where regulations have not been imposed, the situation has gotten worse rather than better. In some populations, 30-40% of asthma cases and 20-30% of all other respiratory diseases have been linked to air pollution. Naturally, the greater the concentration of pollutants, the greater has been the damage. For instance, studies in São Paulo, Brazil, have shown that a 75μg/m^3 increase in concentrations of nitrogen dioxide (NO_2) was associated with a 30% increase in deaths from respiratory illness in children under the age of five.[16]

Legislation in the U.S. spurred automakers to reduce auto exhaust emissions. The auto manufacturers did the rest. They were able to more carefully control the amount of fuel modern car engines burn through two methods. The first was through onboard computers. Computers in modern cars monitor the air-to-fuel ratio to make sure it comes as close as possible for a conventional car to the stoichiometric point, the point at which theoretically all fuel would be completely burned using the oxygen taken from the air.

The fuel mixture actually varies from the ideal ratio quite a bit during driving. The engine tends to be more efficient at higher speeds, up to a point. Sometimes the mixture can be lean (more air, less fuel), and other times the mixture can be rich (less air, more fuel). Reducing the amount of unburned fuel was effective because unburned fuel is a major cause of pollution.

The second method uses a **catalytic converter** (Figure 6). It is located toward the back end of the car in front of the muffler and treats the exhaust before it leaves the car. Most modern cars in first world countries have a three-way catalytic converter that reduces the three emissions most responsible for smog: CO, HC, and NOx.

To combat these three types of emissions, catalytic converters use reduction and oxidation catalysts. The exhaust initially goes through the **reduction catalyst,** composed of platinum and rhodium, which rip nitrogen atoms from most of the NOx molecules, resulting in mostly nonpolluting nitrogen and oxygen molecules.[17]

Next, the exhaust comes in contact with the **oxidation catalyst**, composed of platinum and palladium, which reduces the amount of harmful HC and CO by changing them into water and carbon dioxide.[18]

Three-Way Catalytic Converter

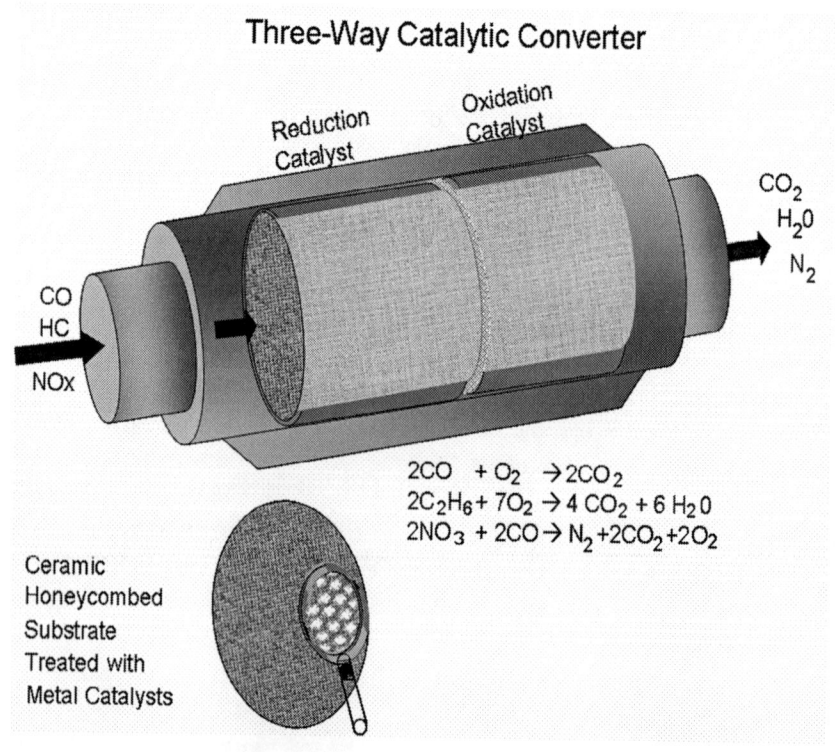

$$2CO + O_2 \rightarrow 2CO_2$$
$$2C_2H_6 + 7O_2 \rightarrow 4CO_2 + 6H_2O$$
$$2NO_3 + 2CO \rightarrow N_2 + 2CO_2 + 2O_2$$

Ceramic Honeycombed Substrate Treated with Metal Catalysts

Figure 6. The Catalytic Converter [19]

Two oxygen sensors, one positioned at the front end of the three-way catalytic converter and the other at the back end, sniff the exhaust and feed status to an onboard computer. Exhaust measurements become data used by the onboard computer to regulate the amount of oxygen (in the air-to-fuel ratio) provided to the engine. Using this feedback control system helps keep the engine running closer to the stoichiometric point while making sure enough oxygen exists in the exhaust so that the oxidation catalysts can burn the unburned hydrocarbons.

While these solutions have made a significant difference, they don't get rid of all the carbon monoxide, hydrocarbons, or nitrous oxides, nor do they remove the other pollutants. Furthermore, fiddling with cars to ensure that their engines run closer to the stoichiometric point for gasoline can go only so far. With the increasing population, the pollution levels, even for controlled pollutants, are beginning to climb again and will pose an increasing burden if we continue to produce conventional gasoline cars.

The pollution problems with the conventional car are unfortunately not limited to the pollution resulting from unburned gasoline or from the use of transmission fluid, lubricating oil, power steering fluid, or coolants. Conventional cars have harbored even more chemicals that can harm the environment or threaten health or both.

The Tetra-Ethyl Lead Man

One of the most significant pollutants not yet mentioned, tetra-ethyl lead, came from the mind of Thomas Midgley Jr., an oil chemist. It helped earn Midgley the dubious distinction of being, according to one historian, the one living organism on Earth that has "had more impact on the atmosphere than any other . . . in Earth history."[20]

Tetra-ethyl lead belongs with a line of additives meant to improve engine performance. Methyl tertiary-butyl ether (MTBE) is a more recent example. Many of these additives have health consequences that were known before they were used. Leaded gasoline could have been counted among the great environmental disasters of the 20th century, had its production and distribution not been phased out where it was most used, in the U.S. The corporations involved in its spread opted for easy profits, confident that the public would buy leaded gasoline to improve car performance and either wouldn't know or wouldn't care about the health concerns.

The main problem that tetra-ethyl lead was meant to address was knock, which is caused by premature detonations within each cylinder. Knock wastes fuel and reduces power by reducing pressure within the cylinders. At the time knock became a critical problem, car companies were trying to make the internal combustion engine more powerful without completely redesigning it. They soon decided that high compression – more pressure – was the answer. The higher the compression the more explosive the fuel mixture would be and the more powerful would be the downward thrust of the pistons, but knock blocked progress.

From 1916 to 1921, chemists at the General Motors research lab in Dayton, Ohio spearheaded the searched for a solution. They tried hundreds of additives and found quite a few possibilities. Ethanol (ethyl alcohol) was for many years their preference.

Thomas Midgley, who headed the team, wrote his boss, GM research Vice President Charles Kettering, creator of the electric starter, that ethanol was "the fuel of the future." Midgley argued that because ethanol was made from celluloid materials that exist in living plants, it would be available indefinitely, unlike oil.

But Kettering and the oil industry weren't buying the idea of ethanol either as a substitute for gasoline or as an additive to it. Their refusal is not without some merit. There are questions today about how viable an option ethanol is in the quantities that would be needed to replace gasoline or even to act as an additive.

On orders from his boss, Midgley started working systematically through the periodic table and a complementary list of chemicals. When he reached the tetra-ethyl compounds he began to get results. Tetra-ethyl tin had some effect, but a solution of tetra-ethyl lead stopped engine knock cold. By December 9, 1921, Midgley was certain he had found the answer they were looking for. Unfortunately, the solution was highly explosive and quite poisonous as Midgley personally discovered. Fortunately for him, he recovered.

Daunted by the unsavory aspects of tetra-ethyl lead, Midgley and his assistant T.A. Boyd continued into the summer of 1922 to champion ethanol fuel from vegetable sources, but the decision had been made. Tetra-ethyl lead was comparatively cheap.

There was little question that it would be profitable. It was an economical solution for the public and GM management was proud of it. To them tetra-ethyl lead was another landmark discovery in automotive history. Accordingly, GM began a marketing campaign that included Midgley. Midgley as the discoverer of its anti-knock properties was asked to hold a press conference on its behalf. Inevitably, he was asked about the potentially harmful effects of lead. Prepared, he went so far as to sniff fumes from a container of tetra-ethyl lead and to pour some of the liquid over his hand to prove it was harmless, though he knew it wasn't.

With a growing acceptance of tetra-ethyl lead, Detroit began to produce high compression engines that increased efficiency and power, counting on leaded gasoline to pave the way, which it did. By 1923 GM had joined with Standard Oil of New Jersey to form a partnership which was eventually called the Ethyl Corporation. (It still exists.) Its exclusive purpose was to manufacture the additive. DuPont, which in 1923 was a one-third owner of GM, also became involved. Other companies quickly found a seat on the honey wagon, including Gulf Oil Company, which banking giant Andrew Mellon was heavily invested in. Gulf was given an exclusive contract to sell leaded gasoline in the Southeast United States.

Tetra-ethyl lead became an established commodity and a time bomb. Almost inevitably, problems began to crop up at the refineries where tetra-ethyl lead was added to gasoline. An incident at a New Jersey Standard Oil refinery prompted an investigation that finally brought with it the glare of publicity. Five workers handling tetra-ethyl lead went violently insane. Several others were hospitalized with less severe symptoms of poisoning.

The manufacture of leaded gasoline was denounced by public health experts like Alice Hamilton of Harvard as a menace to public health. The publicity led to calls for an investigation by the Public Health Service. And it was ordered. But in a flagrant conflict of interest, the Surgeon General was asked to head the inquiry. The Surgeon General as the head of the Public Health Service might seem like a logical choice, but he was an appointee of Andrew Mellon, the banking mogul who had become a major investor in Gulf Oil and who also happened to be Secretary of the Treasury.

When the Public Health Service began to hear testimony in 1925, Hamilton was on hand to denounce GM Vice President Charles Kettering as "a murderer" for distributing leaded gasoline. She cited the dreaded occupational hazards of lead throughout centuries of European history and insisted there were alternative methods of producing an anti-knock, higher octane fuel.

It should be remembered that at this stage, atmospheric pollution was not an issue. The objections were focused on refinery worker contamination, which Standard Oil executives asserted could be controlled. Kettering and others, speaking on behalf of GM, Standard Oil of New Jersey, and the Ethyl Corporation, claimed they knew of no chemical alternatives in the paraffin series that gave anti-knock results. Frank Howard of Standard Oil of New Jersey added that leaded gasoline had come "like a gift from heaven."

Not surprisingly, given the Surgeon General's relationship to Mellon, GM's arguments won out. Ironically, no one, not even Hamilton, had bothered to check the Chemical Abstracts that year which clearly showed that GM was patenting a number of alternatives to leaded gasoline – just in case they were needed. It is now known from confidential GM documents that Kettering and others were very worried about competition from viable alternatives, including catalytic cracking, the use of benzene (an even scarier alternative), ethanol, and higher paraffin-derived alcohols such as tertiary butyl alcohol that Kettering claimed would not work.[21] However, these worries had not led them to curtail production of tetra-ethyl lead, just the opposite. They had pushed production schedules to the limit. Whether the production schedules contributed or not, unsafe conditions resulted in 17 other refinery deaths from concentrated tetra-ethyl lead during the 1920s.

Hearings and protests continued for decades while more and more lead poured into the atmosphere. Then Clair Patterson, the noted geophysicist who is famous for conclusively determining the age of the Earth, entered the fray. Patterson became an opponent after he discovered that lead concentrations in the atmosphere were significantly increasing over time. He added credibility to the fight against leaded gas through his scientific evidence of what was happening.

Yet despite Patterson's efforts and those of others, often in the face of significant intimidation, tetra-ethyl lead continued to be added to gasoline in the U.S. until the 1970's, when fortunately, activism and compelling evidence finally won out. Tetra-ethyl lead was phased out in the United States from 1975 to 1986, but it wasn't phased out in Western Europe until the 1990s. And it is yet to be phased out in some parts of the developing world. After decades of dumping lead into the atmosphere, significant quantities of lead still remain,[22] though the amount has dramatically declined in the U.S. by 94% between 1981 and 2000 as the EPA graph below shows.

Lead Emissions, 1981-2000

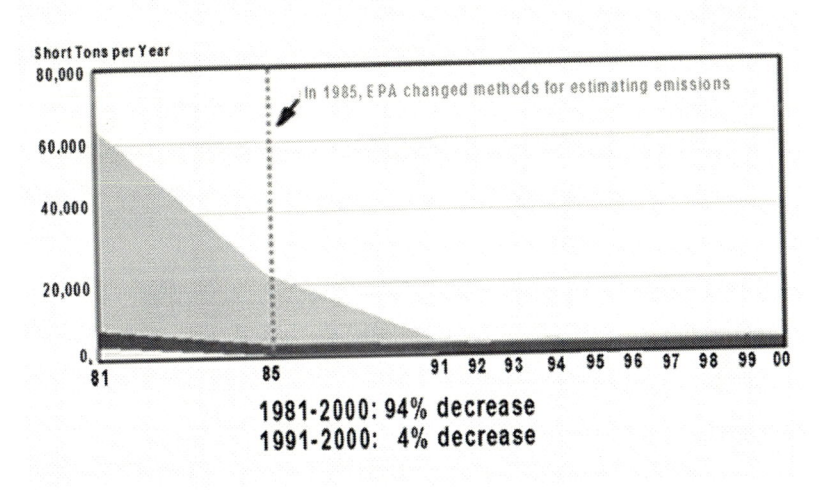

[Graph showing Short Tons per Year decreasing from ~70,000 in 1981 to near 0 by 2000. Annotation: "In 1985, EPA changed methods for estimating emissions"]

1981-2000: 94% decrease
1991-2000: 4% decrease

Figure 7. Lead Emissions in the U.S. from 1981 to 2000[23]

Leaded gas was not, it turned out, Midgley's sole contribution to the degradation of Earth's atmosphere. He made another that could be considered even worse, dichloroflouromethane. Unlike leaded gasoline, dichloroflouromethane appeared to be relatively innocuous – at first. It wasn't directly harmful to human beings. Its use quickly spread, finding its way into everything from automobile air conditioners to hair spray cans to refrigerators. It took years to discover that dichloroflouromethane damages the fragile stratospheric ozone layer, a layer high up in our atmosphere. Unlike the ground ozone layer that can hang harmfully over cities, stratospheric ozone is beneficial. It protects all life on Earth from deadly levels of solar radiation. Dichloroflouromethane attacks the stratospheric ozone layer with a ferocity that is frightening. One pound of the stuff can destroy ten thousand pounds of stratospheric ozone. Like leaded gasoline, it was eventually phased out and its use banned in this country, but it is still being produced.

Thomas Midgley's penchant for the regrettable extended into his own life. Sadly, at age 51, he contracted polio. After he had effectively lost control of his legs, he invented a harness of pulleys and cables to turn him over and raise him in bed. On November 2, 1944, he got tangled in his invention and was strangled to death. He died not knowing the full consequences or fate of his two major contributions to history.

The leaded gasoline story is one of several cautionary tales. The behavior exhibited by the government, oil producers, and automobile manufacturers was reprehensible. They have shown how easy it is for industry to put short-term profit ahead of common sense and ultimately ahead of long-term self-interest.[24] If our government found it so difficult to remove one of history's best known poisons from inadvertent public consumption, how much harder will it be for it to meet the more difficult environmental challenges of the twenty-first century?

On the other hand, the U.S. Government did eventually order the phase out of leaded gas as it has of DDT and dichloroflouromethane, which are hopeful precedents. In this and in many other instances, the U.S. EPA has acted commendably, albeit slowly, to ban harmful substances. What the EPA needs is active support and prodding from knowledgeable, concerned citizens with convincing evidence and logical, reasonable conclusions.

The End of the Hydrocarbon Life-Cycle

Throughout the hydrocarbon life-cycle we see a trail of pollution. At the end of the life-cycle there are only waste products – discarded lubricating oil from car engine sumps, discarded transmission and power steering fluids, discarded car engine coolant, and more. Where does it go? Too much of this waste is still regularly dumped into rivers, lakes, and eventually into the oceans, the ultimate pollution sink. So in the oceans and rivers of our planet, in the sands of our beaches, and in the soil, these poisons commingle with the other unnatural hydrocarbon wastes of modern civilization: the fertilizers, plastics, and pesticides.[25]

Even if we start now, we will still have to deal with the accumulated pollution that lingers in the environment and in our bodies for decades if not centuries to come. Just how much we will have to deal with is still uncertain. But there are hints.

Our Tainted Blood

A study led by Mount Sinai School of Medicine in New York, found an average of 91 industrial compounds, air pollutants, and other chemicals in the blood and urine of nine volunteers. These are substances that should not be there.[26] A total of 167 chemicals were found in the blood of the volunteers if the group is taken as a whole. Like most of us, the people tested do not work with chemicals on the job and do not live near a chemical factory or refinery.

Scientists refer to this contamination as a person's body burden. Of the 167 chemicals found, 76 cause cancers in humans or animals, 94 are toxic to the brain and nervous system, and 79 cause birth defects or abnormal development. There is overlap with some chemicals having more than one adverse effect.

Overwhelmingly, these chemicals are petroleum derivatives, residues, byproducts, or non-oil contaminants such as lead and mercury that are commonly found in petroleum. The contaminants include:

Dioxins and furans — These are carcinogens that are particularly toxic to a child's developing endocrine (hormone) system.[27] They are created and released into the environment as the by-products of human activities. Sources of dioxins are

- The burning of household petroleum based waste (e.g., plastic)
- The production of iron and steel
- The burning of fuel, including gasoline and diesel fuel
- The burning of fuel for agricultural purposes and home-heating
- The generation of electrical power using combusted coal and oil

Organochlorine insecticides — These insecticides are known to cause cancer and numerous reproductive defects. Banned here, the best-known organochlorine insecticide DDT is still produced in other countries and can find its way back into the United States.[28]

Phthalates — Phthalates are widely used industrial compounds known technically as dialkyl or alkylaryl esters of 1,2-benzenedicarboxylic acids. As the name suggests, they contain benzene. They are known to cause not only cancer but birth defects in male reproductive organs.[29]

Volatile and semi-volatile organic chemicals — These are industrial solvents, gasoline additives, and similar products, most but not all hydrocarbon based or closely associated with hydrocarbons. Examples are xylene and ethyl benzene. These benzene products are toxic to the nervous system as well as being carcinogens.

Most of these contaminants persist for decades after they enter the environment. What makes this situation even scarier is that chemical companies are not required to monitor where their chemical products end up. Comprehensive studies are needed to determine both the effects and amounts of contaminants getting into the eco-system, into our food supply, and into our bodies.[30] A comprehensive, multi-state testing program to check human blood for pollutants would be a good start.

What worries me most about our exposure is the possibility that the accumulation of pollutants in our blood has a threshold effect. A slow accumulation may hide its abnormality within normality until an unseen point is reached when a sudden cataclysmic event such as a cancer occurs. The concern extends to the environment. In recent years we've seen mysterious die-offs of animals as diverse as frogs and honey bees. These may be due to threshold effects. Thresholds exist in many organic and non-organic systems. For example, trees growing in an area of heavy acid rain may appear normal until a certain level of acidity is reached in the soil, at which point the compromised trees can no longer compensate for the hostile environment or ward off the predations of their natural enemies and suddenly begin to die off. The ecological consequences of poisoning our environment may well be accumulating to relatively sudden, devastating effect for life on Earth.

Mad as a Hatter

Alice in Wonderland produced a character whose irresponsible behavior reminds me of our own. The Mad Hatter loved to spend time drinking tea and eating bread, but could not find time to clean up his dirty dishes and spills. Instead of cleaning up, his solution was to leave his mess behind and move to another, cleaner spot on the table. As he put it, he moved around "as things got used up." The Earth is not a very big place for billions of people. Its resources are not inexhaustible. Its ability to recover from the pollution we pour into it has limits. By polluting our planet we are ultimately poisoning ourselves. That may have been difficult to foresee back when we began building an energy infrastructure that changed America into a modern society, but today the dark side of what we have created should be all too clear.

The Mad Hatter may not have been simply an imaginary character, by the way. "Mad as a hatter" was a phrase in circulation in the nineteenth century. Today, it is believed hatters were really driven mad by mercury poisoning from a pollutant mercurous nitrate that they used in the manufacture of felt hats.

7
The Climate Threshold

As the amount of pollution we create accumulates, the possibility of reaching a dangerous threshold exists not only within our bodies, in our surrounding ground environment, in rivers, lakes, and oceans, but also in the atmosphere. According to the National Academy of Sciences, since the late 19th century, the global mean surface temperature of the Earth has increased 0.5-1.0°F.[1] Notably, this perceptible rise in mean temperature has accelerated over the past two decades. The 10 warmest years of the twentieth century all occurred in the last 15 years of the century, 1998 being the warmest year.[2] The twenty-first century continued the trend. 2005 turned out to be even hotter than 1998.[3] These are facts. What has been disputed is whether this rise in temperature holds any particular dangers, and whether it is due to an unnatural, manmade build up of greenhouse gases in the atmosphere.

Evidence and Conjecture

The argument that emissions of greenhouse gases by humans are the deciding factor in global warming relies mainly on what has happened in the last 150 years. We know that greenhouse gases are part of a natural carbon cycle. Atmospheric greenhouse gases (water vapor, carbon dioxide, and other gases) trap some of the sun's energy that has reached the Earth's atmosphere. Thanks to greenhouse gases manufactured largely by plants and animals, the Earth's average temperature has been a hospitable 60°F.

However, the atmospheric concentration of greenhouse gases has increased since the beginning of the Industrial Revolution.[4] Atmospheric concentrations of carbon dioxide have increased nearly 30%. Methane concentrations have more than doubled. And nitrous oxide concentrations have risen by about 15%. Carbon dioxide is the trademark greenhouse gas but others are more problematic. For instance, methane traps more than 21 times as much heat per molecule as carbon dioxide, and nitrous oxide is 270 times more potent as a greenhouse gas than carbon dioxide.

What can account for this significant increase in greenhouse gases other than the increased hydrocarbon use that has accompanied industrialization and increased agricultural production? Animal and plant respiration, fires, volcanic eruptions, and decomposition of organic matter are believed to release many more times the CO_2 released by human activities; but these releases, it is argued, have generally been in balance for centuries and form part of the natural carbon cycle in which carbon dioxide is absorbed by terrestrial vegetation and the oceans.

On the other hand, we know that an increase of greenhouse gases has coincided with the Industrial Revolution and dramatic population growth. We know our civilization emits these gases, and we know that they trap heat. So it is not unreasonable to suppose that human activity is to blame. In a 2001 assessment report, the Intergovernmental Panel on Climate Change (IPCC), a UN panel, wrote, "There is new and stronger evidence that most of the warming observed over the last 50 years is attributable to human activities."[5] In 2007 IPCC came out with an even more definitive statement, pointing the finger at human activity.

It is now possible to more accurately measure what humans are dumping into the atmosphere, so the issue of whether humans contribute significantly to global warming is going to be less and less in doubt. Already data are accumulating.[6] For example, in the United States in 1995, 6.6 tons of greenhouse gases were emitted per person. Emissions per person have increased steadily since then. Eighty-two percent of these emissions are from burning fossil fuels to generate electricity and to power our cars. Much of the remaining human-associated emissions are from wastes in our landfills, raising livestock, natural gas pipeline leakage, and industrial chemical plant leakage.

Based on such accumulating data, it is possible to predict future increases in greenhouse gases from current levels due solely to human activity. Several emission scenarios have been developed based on the projections. These can be checked against what actually happens, if we care to wait and accept the consequences. For example, by 2100, in the absence of emission control policies, carbon dioxide concentrations are projected to be 30-150% higher than today's levels. A 30% rise is the rock bottom, most conservative estimate. It doesn't take into account the increased industrial activity in Asia, particularly in China and India.

Besides an increase in the global mean surface temperature, scientists have measured a decrease in the snow cover in the Northern Hemisphere, a decrease in the size of glaciers in North America and elsewhere around the world, a decrease in the size of the ice shelves along the coast of Antarctica, a decrease in permafrost in the Arctic, and an increase in sea level over the past century and a half. Nowhere is the evidence of global warming more dramatic than in the Arctic, where warming temperatures and their effects are most visible. Satellite photos of a shrinking Arctic ice pack are irrefutable.

What will happen next is less certain, but the consensus opinion of the IPCC is not reassuring. They believe that as the greenhouse gases continue to rise, the average global temperatures will rise with them. How much and how fast is still a matter for conjecture.[7] This is essentially the prevailing view of global warming. But there is another side.

The scientists who oppose this view don't necessarily dispute that the Earth is warming or even that the temperature has been rising since the Industrial Revolution. But they argue that underlying assumptions are being fostered that are incorrect.

One assumption that is questioned is that we can continue using fossil fuels at our current rate or greater. That assumption would be incorrect if we are depleting fossil fuel resources at a rate that will soon result in less use of fossil fuels as they become more scarce and expensive.

Another assumption that is questioned is that the average global temperature was not rising before the Industrial Revolution began. In fact, the opposition says, there is strong evidence that the global temperature has been rising since the great chill of the "little ice age," over four hundred years ago, well before the beginning of the Industrial Revolution.

Similarly, they don't dispute that there is a decrease in the ice pack in the Arctic, but attribute it to a different cause than atmospheric warming. It is due, the opposing side asserts, to an increase in warm ocean water entering the Arctic region for reasons that are currently unknown. Along the same lines, the rise in sea levels is not disputed, but these scientists claim that predictions of how high sea levels would rise due to global warming have been too high. And while they may concede that humans contribute modestly to the current global warming trend we are now seeing, they will argue that the major contribution comes from natural forces that are still poorly understood.

In general, the scientists who oppose the idea that humans bear chief responsibility for global warming point out that the Earth has warmed and cooled and warmed repeatedly, and was doing so before humans made an appearance on the Earth. Some suggest that irregularities in Earth's orbit better explain this periodic warming and cooling, but scientific proof of orbital irregularities as a cause of global warming over the last one hundred and fifty years has not been found. Those who believe global warming is largely a natural phenomenon also tend to believe that it will have more benefits than detriments, while those who see humans as the main cause of global warming today tend to warn of dire consequences.

Why We Should Care

Why should we care? Unfortunately, we can't afford to be wrong. At a minimum, if the warnings are correct, global warming will be largely destructive. It will profoundly affect the planet's weather patterns, increasing the severity of storms and drought. And it threatens to precipitate a rise in ocean levels that can inundate coastlines around the world. Even more disturbing, global warming could increase the prospects of an anoxic ocean environment – an environment lacking oxygen.[8] That, together with an increasing oceanic acidity from greenhouse gases, could kill off much of the ocean's biologically diverse life, leaving anaerobic organisms to multiply. The anaerobic organisms that are likely to multiply in an anoxic ocean environment are known to produce waste products, mainly noxious gases, that are poisonous to aerobic (oxygen breathing) life. If these anaerobic organisms spread uncontrolled, they could kill off most life on Earth and what remains of aerobic life in the oceans.[9]

There is a further, possibly more imminent danger. Global warming could cause the release of large quantities of stored methane hydrate into the atmosphere. Methane hydrate is found in large quantities in certain oceanic and polar regions. It has been bottled up in sediments for millennia thanks to cool temperatures. A sudden, large, and sustained release of methane into the atmosphere could actually accelerate global warming with devastating consequences because it is over 21 times more potent as a green house gas than carbon dioxide. The most likely mechanism of release is Arctic melting or underwater landslides, but there could be other mechanisms, including human tampering. Methane hydrate has been touted as the next energy source once oil is depleted. Arctic melting, however, remains the more likely threat. The release of large quantities of Arctic methane hydrate into the atmosphere could be only a few decades away.

Bolstering fears is a new theory that a release of greenhouse gases has occurred before. It isn't good news, if true. According to the theory, at the end of the Permian Period 250 million years ago, methane hydrate along with gases from anaerobic bacteria proliferating in warming anoxic oceans and outgassing from erupting super volcanoes combined to cause a catastrophe. The Permian-Triassic boundary saw the greatest extinction of life in Earth's history.[10] Over 95% of all marine life and 70% of all terrestrial life died off.

The great concern of the Intergovernmental Panel on Climate Change (IPCC) is that an unnatural, manmade increase in greenhouse gases will have an unexpected non-linear effect. A non-linear effect could occur if a threshold trigger event is reached, and Earth's compensatory mechanisms are overwhelmed. The catastrophe that ended the Permian period might have been an example of such an effect. In 1995 IPCC cautioned, "Complex systems, such as the climate system, can respond in non-linear ways and produce surprises." Steve Schneider, a distinguished member of the IPCC, concluded in the panel's 1995 assessment report, "My free translation of this concern is that reducing the pressure that humans put on nature is an insurance policy against 'nasty surprises.'"[11]

Unfortunately, the United States emits more greenhouse gases per person than any other country. If human actions are a major contributing factor to global warming, we bear more responsibility for what is occurring than any other nation. Our only rival currently is China, which in overall pollution emissions now surpasses us. Because we provide so much of the world's pollution, we have a major responsibility to ourselves and to the rest of the world to do something. Waiting for unequivocal proof that we are adversely influencing our climate is a dangerous position to take.

By the time science can provide us with irrefutable evidence of what is actually happening and why, we could face a crisis of such proportions that we can do nothing to stop it. To ensure our actions don't become the deciding factor in the ill health of our planet, it seems far better to "reduce the pressure that humans put on nature." It makes more sense to take out an insurance policy against "nasty surprises" and never know what might have happened if we had not, than to wait and potentially verify our worst fears.

8

Population Growth Blow Out

It is not simply the unstable situation in the Middle East or the pollution we currently tolerate that should convince us that we must find alternatives to hydrocarbons. No one has a crystal ball, but it doesn't take much thinking to realize that rapid industrialization founded on hydrocarbons and explosive population growth can only magnify existing dangers. We are seeing a convergence of dangerous trends right now.

The Increase in Instability

In the most unstable countries of the Middle East, and in Asia and parts of sub-Saharan Africa, the population is not only growing larger, it is growing younger. The majority in countries like Iran and Egypt are under the age of 25.[1] With the rise in a youthful population has come a rise in volatility. Most of the societies in which population growth is out of control are the least likely to provide general prosperity or fulfill the aspirations of their young people. These societies are breeding unrest among the most vigorous and most politically naive of their population through unemployment, physical abuse, neglect, and stagnation.

Disruption and Disorientation

Much of the youthful population in third world countries is torn between a desire for industrialization, which is bound to be culturally disruptive and disorienting, and the call of elders to observe traditional values that make change difficult if not impossible. The young are caught between two worlds.

Exacerbating third world problems, rigidly patriarchal customs within many of these societies continue to foster the out-of-control population growth and impoverishment. These customs have hardly changed in two thousand years, but their impact has never been more detrimental.

In many of the poorest countries, marriages are still traditionally arranged. Usually the bride and groom barely know one another if at all before the wedding. For the young bride, it can be a rude transition, even if she marries within her own tribe. Forsaken by their own families, brides are forced to enter a family that often has little sympathy for her and in which her powerlessness can be overwhelming. Both the female and male relatives of the husband have a vested interest in the bride having male children because male children are the future economic providers of the extended family. The young woman will be forced to have large families in order to ensure that a sufficient number of male children are born. Her new relatives will look down upon her if she has female children, perpetuating an attitude she is likely to adopt herself. Just as male children are a boon to a family, female children are a burden because the family must pay a dowry and because female children belong ultimately to another man's family. Female infanticide under these circumstances has often been an option. But even today, it has not significantly reduced the tendency to sustain six or more children per family, a recipe for increasing impoverishment and violence.

Rapid Industrialization

Where modernity has been embraced rather than rejected, and women have been empowered and educated, as in the West, populations have been stabilizing or even declining. But modernity brings problems as well as benefits, especially for countries with immature economies. In China, which now straddles East and West, old and new, the unexpected embrace of Capitalism by the Communist regime has certainly led to an unprecedented industrial revolution and increasing prosperity. But the increasing industrialization has been accompanied by severe environmental problems. While China has managed to curb population growth, that feat, from an environmental perspective, has been offset by rapid industrialization which has increased pollution. China still depends more on combusted coal than on oil and natural gas for cooking and heating, which has made China the home of seven of the ten most polluted cities in the world.

As it seeks to expand its huge economy, oil-poor China is becoming a major importer of oil, exerting greater and greater pressure on the known reserves of oil all over the world and increasing competition for what must at some point become a shrinking supply. Therefore, even though its population growth is slowing, China's growing economic prosperity, if based on petroleum, ensures that the competition for oil will grow more intense in the future, and China's sizable coal reserves ensure that pollution around the world will increase dramatically.

With the industrialization of China and India, with the continued explosion in population, with the contention between ideologies exacerbated by industrialization, with all of this tied to a growing consumption of hydrocarbons, we are treading dangerous waters. We are already experiencing a significant increase in world instability that could lead to an overt third world war sparked by contention over dwindling oil supplies. If we are going to mitigate this risk or at least prevent our country from being dragged into a vortex of conflict over resources, we simply have to reduce our dependency on foreign oil.[2]

9

The Hidden Costs of Oil

The economic, political, and social dangers of our dependency on hydrocarbons have been clear since at least the 1970s. Yet over thirty years after the 1973 Arab oil embargo, the United States is more vulnerable than ever. In 1972, the year before the embargo, the U.S. imported 27.6% of the oil it consumed. Today, we import over 60% of our oil, more than twice the 1972 level. Even a modest disruption of these imports could be grave. Yet from the 1990s until recently, most consumers didn't seem that concerned or even aware of the danger, judging by our preference for gas guzzling cars. Why? My guess is that most of us use the price we pay for gasoline as the measure of whether we should be concerned or not. So long as gasoline appears cheap, oil seems abundant, and the risks are ignored. The tendency to ignore the risks of dependency is buttressed by the belief some hold that our domestic oil production can still be easily stepped up at any time to meet our needs. Without correct information, we come to the wrong conclusions.

At least the price Americans have been paying at the pump has finally gotten everyone's attention. But reacting only when prices for gasoline rise to say $3.00 per gallon or oil reaches $90 a barrel is a little like deciding to wait until the fuel gauge indicates empty, before trying to find a place to refuel. That may seem foolish, but it is an apt analogy. Consider the trend in gasoline prices between 2003 and 2007, adjusted for inflation. In 2003 we paid on average $1.52 per gallon; in 2004, $1.79; in 2005, $2.28; in 2006, $3.03; and in 2007 we paid $3.26. During this same period, the OPEC price of crude oil shot up from under $25 a barrel to over $95 a barrel.[1] The trend was clear despite the price fluctuations, yet when did we begin to react?

If price rises are late indicators of problems, price decreases are even less help in judging what we face. In fact, they can mask dangerous problems. In the 1990s we had some of the lowest prices for oil in 20 years. During the same time our dependency on foreign oil quietly increased, leaving us open to the dramatic price rises we see today.

What's more, even at its lowest point in the late 1990s, the "low price" we paid then for gasoline was higher than it seemed. Consider the obvious costs of producing and distributing petroleum products such as gasoline:

- oil exploration
- drilling rights for oil from foreign countries
- set up and maintenance of an extraction system
- labor and materials to actually extract the oil
- transportation of the oil
- labor and materials to build refineries
- processing of the gasoline at a refinery
- transportation of the gasoline to a dispensing point
- distribution to customers

Given everything it takes to create the oil products we consume, the price of gasoline might seem like a bargain, but what hasn't been taken into account are the external costs of oil. These are costs that work their way through our economy largely unnoticed, like creeping economic pollution. Some of these external costs are difficult to calculate precisely, others can be found readily with a little digging. Several organizations and individuals have sought to make estimates based on existing data.

A report by the International Center for Technology Assessment (CTA) back in 1998 estimated that when the external costs are added to the retail price, the real cost of gasoline is $5.60-15.14 per gallon.[2] The writers arrived at their estimate after looking at cost categories such as:

- Government tax subsidies to the oil industry
- Security costs [3]
- Environmental and health costs of gasoline usage

According to the National Defense Council Foundation (NDCF)[4] in a more recent 2003 study, the real price of gasoline was around $5.28 per gallon. The more conservative National Defense Council Foundation looked at the costs related to:

- Government tax subsidies to the oil industry
- Security, especially military expenditures defending Persian Gulf oil
- Transfer of wealth
- Economic shocks

These two sample estimates don't agree quantitatively because the problem is not timeframe so much as reference. Organizations that have looked at oil costs haven't worked from exactly the same list of cost categories or specific expenses. It's not surprising then that the figures quantitatively differ. But the important conclusion that their results agree on is that our addiction to oil is costing us much more than just the price at the pump.

The real costs become more obvious when the hidden costs enumerated by various organizations are consolidated into a reasonably comprehensive list of cost categories like this one:

Tax Breaks and Subsidies — Costs of direct and indirect public subsidies to oil companies, including tax breaks enjoyed by American oil companies.

Security — The increasing costs of protecting and securing foreign sources of oil as well as the costs of protecting and securing pipelines and refineries or any other part of the oil infrastructure that may be subject to attack at home or abroad.

Health and Environment — Costs of human and environmental health problems caused by air, water, and soil contamination by hydrocarbons. This category includes health costs to the general public, health costs incurred by workers in hydrocarbon related industries, and costs of environmental damage. The latter costs encompass the costs of pollution in the air, the water, and in the soil.

Transfer of Wealth, Job Loss, and Economic Shocks — This cost category reflects the impact of importing oil on the stock market, the job market, the national debt, and more generally on the price of goods and services. Money that goes to foreign countries is money not spent on businesses and industries here. Some sectors of the economy, like the airline and trucking industries, will suffer more than others. These most vulnerable industries, like canaries in a mine, provide a forewarning of economic hard times. Their problems seldom remain isolated. They usually pass on their costs, which can produce a ripple effect throughout the economy.

While it accounts for most of the external costs, this list doesn't include everything. It doesn't take into account the pain of a mother and father whose son or daughter has died protecting the Persian Gulf oil region or has died in terrorist attacks financed directly or indirectly by oil money. It doesn't attempt to estimate the potential cost to Western countries if oil production were drastically reduced due to the coordinated sabotage of oil fields, tankers, and pipelines. Nor does it calculate the future costs of competition or the cost of potential conflict with China or other countries over dwindling oil reserves in the Middle East and elsewhere. Yet each cost category listed, if viewed in more detail, can provide a more realistic idea than we generally have of what we really spend for oil.

Tax Breaks and Subsidies

Federal tax breaks and subsidies that directly benefit oil companies are a matter of record. Researchers at the International Center for Technology Assessment (CTA) dug into the record and came up with figures in their report for 1998 when the price of a barrel of oil was at an apparent low point. Here is some of what they found:

- Percentage Depletion Allowance Subsidy – $0.78-1 billion
- Non-conventional Fuel Production Credit – $769-900 million
- Exploration and Development Costs Subsidy – $200-255 million
- Enhanced Oil Recovery Credit Subsidy – $26-100 million
- Tax Break: Foreign Tax Credits – $1.1-3.4 billion
- Accelerated Depreciation Allowances Subsidy – $1.0-4.5 billion

In total, CTA calculated that the annual tax breaks and subsidies were between $9.1 and $17.8 billion. The National Defense Council Foundation (NDCF) looked into the tax breaks and subsidies five years later in 2003 and estimated lost federal and state annual revenues at $13.4 billion.

Tax subsidies and tax breaks do not end at the federal level. The fact that most state income taxes are based on oil firms' deflated federal tax bill results in an estimated under-taxation of $125-323 million per year. The subsidies and tax breaks have continued despite huge profits for the oil companies in 2006 and 2007.

Tax Breaks and Subsidies: Another Viewpoint

A view that supports tax breaks and subsidies is provided by John L. Moore, Carl E. Behrens, and John E. Blodgett in a 1997 Congressional Research Service (CRS) report for Congress entitled *Oil Imports: An Overview and Update of Economic and Security Effects*.[5] Although this report, like the CTA report, was written when gasoline prices were unusually low for that period, it provides insight into thinking that continues in some political circles today. While conditions change, perceptions often lag dangerously behind.

In their report, the CRS authors describe the oil industry tax breaks and subsidies as risk mitigations – actions that have helped reduce potential risks. Behrens and Blodgett don't mention the cost of subsidies but assert that subsidies promote "an efficient and competitive domestic oil industry through policies on oil and gas leasing, research and development, and equitable tax treatment."[6] They suggest that the price of domestic oil can influence the global price, keeping it lower.

Security

Besides the subsidies, there are other costs of oil products usually not taken into account. One is security. The costs of protecting the oil far exceed the amount paid out in subsidies and tax breaks, according to both the International Center for Technology Assessment (CTA) and National Defense Council Foundation (NDCF). They offer different estimates, but mainly because they use different sources for their estimates. Here are their breakdowns.

The CTA estimates that the Defense Department spends $55-96.3 billion per year to safeguard the world's petroleum resources. Most of that money goes to prop up and protect Saudi Arabia and its neighbors.

The NDCF estimate is significantly lower because the NDCF bases its estimate on the budget of the Central Command, which has a mandate to defend Persian Gulf oil. NDCF cites a cost of roughly $42.8 billion for this specific mandate. NDCF adds to this cost the Central Command's annual contingency expenditure for protecting the Persian Gulf in order to arrive at a cost of approximately $49.1 billion.

CTA includes costs not considered by NDCF. Examples are

- Protection Services of the Coast Guard
- Local and State Protection Services
- Strategic Petroleum Reserve

(The Strategic Petroleum Reserve is a federal government entity designed to supplement regular oil supplies in the event of disruptions.)

Neither organization included expenditures incurred in the first Gulf (Iraq) war. Nor could they have foreseen the second Gulf (Iraq) War and its costs.[7] Some may argue that the wars in Iraq had nothing to do with oil and that they are rather part of the war on terrorism. I have already expressed my skepticism. It's hard to believe Saddam was not after his neighbors' oil rich regions, given how much their oil would have increased his wealth, power, influence, and prestige. It is also hard to believe that the decision to counter him had nothing to do with oil.

Security: Another Viewpoint

The authors of the 1997 Congressional Research Service (CRS) report for Congress on oil imports hedged on the cost of providing security for oil. Quoting an earlier CRS 1992 report, they stated that the security cost of Gulf oil imports "is either insignificant or ponderous, depending on the assumptions made." They presented various costing assumptions:

> "If only those military activities are tallied which would not be undertaken in the absence of oil in the region, then the cost of defending the oil would be very small. If any military activity is counted that, undertaken for whatever reason, contributes to the security of the region, the cost calculation becomes much larger... Some analysts used the relatively small proportion of oil imported to the United States from the Gulf countries as the base for calculating a 'per-barrel security cost' of import dependence. Others used the entire production from the region. Still others pointed out that ...an interruption of cheap Middle East oil supply would push up the price of oil for everyone. From that point of view, the proper denominator is world oil production, leading to a very small per-barrel figure."[8]

They concluded, "There may be other reasons for supporting the effort to reduce imports, but they will have little effect on the amount spent for security in the Persian Gulf."[9] The report did attempt to make an estimate of security costs, which had a wide range of $0.5-65 billion annually.

Health and Environmental Costs

The International Center for Technology Assessment (CTA) 1998 report cited environmental, health, and social costs as a large portion of the unreported price that Americans pay for their gasoline. It estimated annual health costs from automobile pollution at $29-542 billion and pollution damage to buildings at another $1.2-9.6 billion.

Health and Environmental Costs: Another Viewpoint

The 1997 CRS report acknowledged that there are health costs but described a "reasonable estimate" of the annual health cost of pollution to be around $4 billion.[10]

Until a definitive study can be done that is similar in scope to the studies done to investigate a link between smoking and cancer,[11] the health consequences and costs of hydrocarbons and associated contaminants[12] may be hotly disputed.[13]

Job Loss, Supply Disruptions, and Economic Shocks

In its 2003 analysis of the hidden cost of imported oil, the National Defense Council Foundation (NDCF) stated,

> "Relying on overseas sources for oil has sent an average of 138,036 jobs overseas annually and has cost the U.S. economy another 345,090 jobs from secondary employment that would have been generated by the lost domestic economic activity. The loss of employment, in turn, has robbed the U.S. worker of an average of $20.5 billion in wages annually and has caused the cumulative loss of some $212.3 billion in state, federal, and local taxes."[14]

The Department of Energy (DOE) estimates that each $1 billion of trade deficit costs America 27,000 jobs.[15] Our oil imports account for around one third of our annual deficit, a percentage that is rising. Therefore, oil imports are a major contributor to unemployment. NDCF estimated a total economic penalty of importing oil from $297-305 billion annually,[16] a staggering number. Since 2004 this situation has worsened.

Why is the impact so severe? Part of the reason given is that when the OPEC cartel drives up prices by restricting supplies, it creates wealth for itself without the need to expend real resources; that is, by disciplined production cuts the cartel is able to appropriate wealth from the rest of the world with little or no reciprocation. According to a report issued in 2000 by the Oak Ridge National Laboratory (ORNL), OPEC's members have this power because of their monopolistic control. Our society and much of the rest of the modern world can't function without their oil. We have at present no economical and readily available substitutes. Therefore, OPEC can dominate the world oil market. They have the power to squeeze out competition and control prices everywhere. ORNL observes that OPEC's exercise of market power and extraction of monopoly rents would not be possible in a more competitive market.[17]

The report by ORNL adds evidence to buttress the premise that oil imports cost jobs and disrupt the economy. Zeroing in on two major oil price spikes between 1970 and 2000 (1979-81, 1999-2000), ORNL observed that each spike was followed by an economic recession in the United States.[18] In each case, there was a lag before the full effect of oil prices kicked in. A recession in 2008 looks like it will follow this pattern.

ORNL clearly sees a connection between oil supply disruptions or price spikes and recessions and makes a persuasive case, but it is not an open and shut case. Other factors must figure in. It's worth noting, for instance, that the increases in the price of oil occurred during a time of upheaval in the Middle East: the Arab war with Israel in 1973, and the Iran-Iraq war begun in 1979. Nonetheless, the oil supply disruptions and spikes in price can hardly be considered coincidence.

The National Defense Council Foundation has a similar finding. It puts a dollar amount on the oil supply disruptions we faced between 1979 and 1981 at $956-1,105 billion.[19] This might be compared with ORNL's estimate of $7 trillion for the cost to the U.S. economy of oil supply disruptions and price spikes between 1970 and 2000.[20]

While political events undoubtedly had a major effect on the price rises during the periods cited above, the subsequent rise in oil prices in 1999 appears to have been influenced mainly by economics. ORNL observes that in 1999 OPEC exercised its dramatic power over the oil market for economic rather than political purposes. In 1998 the oil price had dropped to a level not seen since the 1970s – around $10 a barrel.[21] The low price of oil and the perceived success of the first Gulf War caused many to ignore the bigger picture. But it was soon clear that the price level was not acceptable to OPEC. Understandably so. OPEC in collaboration with Mexico and Norway reduced production. During 1999 prices not only returned to $20, they climbed to $30 per barrel. Unfortunately, even $30 per barrel hardly seems significant now.

Naturally, American oil companies reaped the benefits of this rise, but they did not initiate the price hike. As I've already noted, American oil companies in the Middle East essentially ceded control of the oil fields to the host countries and their cartel decades ago.[22]

Job Loss and Economic Shocks: Another Viewpoint

The 1997 CRS report quotes the previous 1992 CRS report when estimating the effect of oil price shocks.

> "Estimates for this potential component of an oil import premium in the 1992 CRS report were in the $6-9 billion range, based on a variety of studies. Such estimates are based on various macroeconomic energy models using varying assumptions on probabilities and magnitudes of price shocks. The review and analysis by Bohi and Toman notes a Department of Energy estimate from 1991 in the range of about $3-8 billion annually."[23]

More pessimistically, the 1997 CRS report suggests that we have no choice and limited capabilities to avoid oil shocks.

> "Even if the United States could reduce oil imports substantially from politically sensitive regions, the integrated nature of the world oil market means that the domestic economy could still be subjected to potentially wide swings in oil prices."[24]

Furthermore, the CRS report questions whether we should act to reduce imports at the cost of the economy when it will have little impact.

"It is sometimes argued that any reduction in U.S. imports would reduce the 'pressure' on the world market, but that position is conceptually flawed. In the absence of a shortage of a commodity – the proved world oil resources have increased steadily since the 1973 energy crisis – market forces tend to balance demand and supply. Reducing U.S. consumption would reduce world production rate, but it is hard to see how this would increase the stability of the market."[25]

Our Current Economic State and Future Projections

But why should we care, given that our economy seems to tolerate our oil burden? Consider first Figure 8.

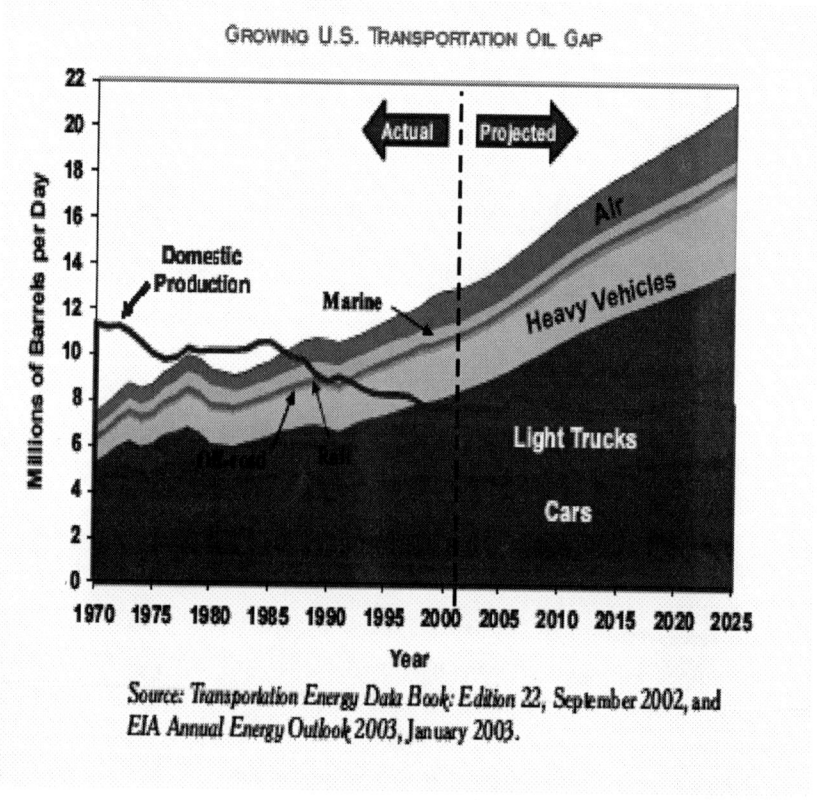

Figure 8. Domestic Oil Production[26]

What the graph shows is that our domestic oil production has declined to a point that it can no longer sustain our needs. The projection to 2025 made in 2002 suggests that domestic production would stabilize at around 8 million barrels a day, but that has not been the case. It has continued to decline. Nor can we simply blame our growing population for our failure to meet our domestic needs. The popularity of SUVs – designated as light trucks in the graph but used as passenger cars – has accelerated our need for oil imports. We have been buying them in record numbers. With their heavy body weight and greater thirst for gasoline,[27] SUVs had by 2005 already reduced the average mpg for "light duty vehicles" which includes SUVs and regular passenger cars to 20.2 mpg, which is below the 1987 average.[28]

While our dependency on oil imports was growing to over 60% of our consumption, the Middle East was gaining some notable new customers in Asia, with the likely consequence that the increasing competition for oil will cause our national deficit to balloon to even more dangerous levels than they have already reached.[29]

Accelerating Oil Consumption

How much competition for oil can we expect? From now to 2025, world oil consumption is projected to increase by 30-50%, according to the U.S. Department of Energy, Energy Information Administration (EIA).[30] Transportation will be the fastest growing oil-consuming sector. Global consumption of gasoline could easily double. U.S. demand is expected to follow but dip below the world trends. In the next two decades, India and China's oil consumption is expected to grow significantly, making it strategically imperative for them to secure their access to oil.

Global population is projected to reach eight billion by 2025 with more than half of those additional people born in Asia and Latin America, while the number of passenger cars will increase to well over 1.1 billion from approximately 550 million today.[31] On a per day basis, global oil consumption is expected to grow from 80 million barrels per day (mb/d) in 2004 to 89 mb/d in 2010 to 103-120 mb/d by 2030. These projections assume that the market can deliver such quantities.[32]

Dangerous Sources

According to the Institute for the Analysis of Global Security (IAGS), six countries in one region of the world (the Middle East) hold 64% of the world's easily accessible oil reserves.[33] (Easily accessible oil reserves are those that can be easily pumped out of the ground.) Here is IAGS's breakdown: Saudi Arabia (25%), Iraq (11%), Iran (8%), UAE (9%), Kuwait (9%), and Libya (2%).[34] British Petroleum (BP) has also estimated the distribution of proven, easily accessible oil reserves in its annual reports "Statistical Review of World Energy."[35] BP's estimates approximately agree with those of IAGS. Estimates from 2005 through 2007 barely changed these numbers.[36]

A Potential Decline in Oil Production

Making the picture even darker is the projected peak and decline in world oil production predicted around 1949 by the late geophysicist Dr. M. King Hubbert. In what has become known as the **Hubbert Peak Theory**, he predicted that we would reach a world oil peak around 2005. According to Hubbert, one signal of the projected peak in world oil production would be a dramatic price leap, after which prices will never move downward again to any significant degree unless consumption falls at a rate commensurate with production.[37] Lending credibility to his world oil peak prediction, Dr. Hubbert correctly predicted the U.S. oil peak in 1970 and the subsequent decline in U.S. oil production.

But Dr. Hubbert's world peak prediction, which can not yet be confirmed, has doubters. When asked about the Hubbert Peak Theory in an interview, Dr. Gal Luft, the executive director of the Institute for the Analysis of Global Security (IAGS),[38] expressed some of the reservations. He pointed out that Hubbert's prediction was particularly accurate with respect to the U.S. because the U.S. has been thoroughly explored for oil. Hubbert had less data on which to make his world oil peak prediction.[39] Luft also argued that Hubbert's world oil peak prediction didn't take into account oil from less easily accessible sources, from the less explored regions of the world, including the Arctic and Antarctic.

Yet many, if not all, of the points Luft makes regarding the status of oil today, buttress Hubbert's theory. Here are Luft's most salient points:

Most oil fields with known reserves are decades old. According to Luft, seventy-five percent of the easily accessible oil produced today comes from less than one percent of all fields. These are the giant fields like Ghawar in Saudi Arabia. Most of these fields are declining. The average age of the world's 14 largest oilfields is more than 43 years.[40]

THE HIDDEN COSTS OF OIL 101

Spare capacity is eroding. Spare capacity is the capacity above the current demand. Few countries can claim to have spare capacity today. One is Saudi Arabia. Luft points out that the Saudis claim they have 2-3 million barrels per day (mb/d) of spare capacity. He adds that credible doubts have been raised about such claims. And Luft notes, "The implications of loss of spare capacity could be severe…if a major producer like Nigeria, Iraq, or Venezuela goes offline due to strike, riots, an embargo, or terror attack there will be nobody to replace it and there will be nothing to stop the price of oil from going through the roof."[41]

New finds of oil aren't keeping pace with increased demand. According to Luft, "Between 1995 and 2002 an average of only about 10 billion barrels a year have been found. … Since 1980 oil production has exceeded new field discoveries by a considerable margin."[42]

Newly found oil is harder to access. The oil that is being found now is generally more inaccessible and harder to extract and process. The best example in the Americas is perhaps the tar sands of Alberta, Canada.[43] Other examples include the Bakken Basin shale oil in the U.S. and Venezuela's mudstones.

The potentially good news, according to some, is that the tar sands of Alberta together with the oil shale in the Bakken Basin contain more oil than the Middle East.[44] If the estimates of oil reserves in Alberta are incorrect, however, and we are indeed close to a world oil peak, our situation could already be dire.

Even with the most optimistic estimates, the news isn't that good. Extracting oil that is difficult to access and process, naturally takes more energy and costs more than extracting oil from oil wells.[45] Even extracting tar sands that are near the surface, which requires the least energy expenditure, has drawbacks according to Luft. First, the extraction process is likely to result in severe groundwater contamination. The contamination would be due mainly to the leakage of materials that dilute the oil in the tar sands. Second, the process requires a lot of natural gas, which is already in tight supply. He doesn't think pegging future U.S. oil supplies to the whims of the natural gas market makes sense.

An Aging Infrastructure. Luft adds to these concerns, worries about our aging oil infrastructure and the shrinking number of refineries in the country. To meet projected demand in 2010, he estimated the oil industry would need to invest up to $1 trillion, which is substantially more than it is spending today.[46]

Other organizations and individuals have made the same or very similar points. For example, Matthew R. Simmons, president of Simmons and Company International, an energy investment bank, has challenged Saudi claims of spare capacity. He believes their figures are not reliable. He doesn't believe their overall capacity will climb much higher than its current 10 mb/d, with no spare capacity. In fact, he believes they are hiding a decline in capacity. Consequently, he doubts Saudi Arabia will be able to satisfy the world's thirst for oil in coming years. Simmons bases his argument on an analysis of over 200 technical papers on Saudi reserves by the Society of Petroleum Engineers. Adding to his credibility is the fact that his work was peer reviewed by a dozen senior technical experts. The Saudis have rushed to refute Simmons' work but have refused to open their books and allow a third party to investigate their reserve data. Simmons has come out with a book defending his claims.[47]

Buttressing these doubts was an article in 2004 in the New York Times that claimed the Ghawar oil field, the largest in the world, has peaked and is declining.[48] The New York Times article describes the signs of a played out field, in particular, the amount of water that has to be pumped into the ground. Water is pumped into oil fields to increase pressure so that oil can be pumped out. As oil fields are depleted of oil, more and more water must be pumped in to maintain pressure. Increasing the amount of water increasingly dilutes the oil that is pumped out. The Ghawar oil field shows the telltale signs of depletion since the Saudis have to pump 7 million gallons of water a day into the oil field to maintain pumping pressure. Saudi Arabia has increased their production in the past to keep the world from having market shortages. The article asked whether that is still possible.[49]

Worries about the aging oil production infrastructure were partially corroborated in reports about BP's Alaska pipeline, which ruptured in March 2006. The rupture was attributed to aging.

A Dangerous Vulnerability

So the price of gasoline at the pump is both a poor and a late barometer of economic trouble, though it appears to be the main driver of action. It is unreliable because, to start with, the price at the pump is not its true price. If hidden costs are taken into account, the cost of gasoline in the U.S. is much higher than the public imagines. But more importantly, it is unreliable because it is too short-sighted a measure. Using the price of gasoline alone, we are missing the broader implications of importing oil.

To recap the problem, despite the heavy subsidies of domestic oil, we are increasing our dependence on imported oil at a time when global world production of easily accessible oil is showing signs that it will be hard pressed to keep up with demand. Most of the largest known oil fields are decades old. Many new oil finds are years away from being productive while the oil appears to be much more difficult and costly to access, extract, and process. We are paying an increasing penalty in health costs, thanks to consumption of hydrocarbons, their derivatives, and associated impurities. We face enormous costs related to the clean up of the environment. And we are likely to spend much of this century grappling with global warming, which we can reasonably assume is exacerbated by the combustion of hydrocarbons. Moreover, with much of the world's readily producible oil reserves in the hands of despotic governments hostile to the United States and Europe, with China and India expanding at an explosive rate and competing for a greater and greater share of the world's easily accessible oil reserves, with Russian oil and natural gas reserves becoming increasingly vital alternatives to Europe, the political, economic, social, and environmental consequences of reacting solely to the price of gasoline at the pump are dangerous, almost beyond imagining.

10

Back to the Future

So far we have not reached a crisis stage. Our economy appears deceptively normal, despite some tremors and a growing recession.

The Risk Zone

We remain in the risk zone where devastating energy crises have not materialized but a multitude of disturbing trends vie for our attention. Because risks are not certainties, we often fail to act but rather wait for a crisis, which has an obvious immediate impact. It will be a tragic mistake to wait for crises before acting this time. Consider two of the dangerous paths we could take from here.

The Downward Spiral

The first scenario that could unfold if we fail to act in time is what I call the "downward spiral." Suppose we continue to ignore the warning signs. Prices of gasoline go down for a while. Other problems intervene to distract us, magnified by our oil supply problems: an increasingly out-of-control trade imbalance, unprecedented levels of personal debt, a loss of jobs to the cheaper Asian labor market, a massive national debt growing even larger because of the war on terror. Eventually, oil supplies begin to dwindle measurably then so quickly that we can't ignore the problem, and we can't adjust to it.

We will have entered an initially slow but increasingly rapid and self-sustaining downward economic spiral. Under this scenario, ever-increasing amounts of money are spent on crude oil and on protecting virtually indefensible oil pipelines, barges, refineries, and foreign reserves until we reach a point at which we can no longer afford the very thing that would stop the downward spiral, domestic alternative energy.

An Energy Melt Down

If our economy is jolted by a sudden steep drop in oil supplies, we could face an even more frightening scenario, an energy melt down not unlike an environmental trophic cascade.[1] In 1973 we suffered long lines after a 7% decline in oil imports. Imagine what would happen if our oil imports suddenly dropped by 18 or 25%? As with the stock market crash in 1929, I believe the initial jolt would send the stock market falling precipitously. Spending on goods and services would dramatically decrease. People would spend hours away from work at gasoline service stations, waiting to fill up as they did in the 1970s. Food distribution would be disrupted. Gasoline rationing would have to be imposed. The price of vital goods would rocket up as the costs of hauling goods by truck, plane, or train soared. Deprived of gasoline, workers would struggle just to get to work. Stores would have fewer and fewer goods on their shelves. Week by week the number of businesses going bankrupt would increase. The political divisions within our society would deepen. The transportation sector would come apart. Beyond our borders the problems would grow in severity as well. In such a scenario, totalitarian demagogues could gain an even more formidable following.

Of the two scenarios, the "melt down" may be more likely than the "downward spiral." How could it happen? You can put the scenario together for yourself from what is already occurring. Add to what you already know, some information that may not have gotten as much attention as it should have. Here are some actual headlines you probably haven't seen or have forgotten:

>Massive Attacks Halt Iraqi Oil Flow[2]
>China, Iran, Russia, and Venezuela are making all kinds of oil deals[3]
>Deadly blast on Nigerian pipeline[4]
>Al-Qaeda 'behind Saudi oil plot'[5]
>China and Venezuela sign oil agreements[6]

Suppose the public demands lower gasoline prices. Suppose we are lulled by temporarily lower gasoline prices and allow our dependency on imported oil to continue increasing. Suppose that meanwhile China makes deals with all the major oil producing countries including Russia, Saudi Arabia, Iran, Libya, Nigeria, Indonesia, and Venezuela. Suppose Iran's scientists, flush with oil money, are able to produce nuclear weapons and spark an arms race in the Middle East or spread the technology to enemies of the U.S. Suppose Pakistan's pro-Western government is overthrown by Islamic extremists. Suppose the U.N. does nothing substantial to stop the spread of nuclear weapons, thanks in part to vetoes from cash hungry Russia and oil hungry China. Then suppose that some Islamic and socialist-Communist oil producing countries form an alliance and decide to embargo oil to the U.S., convinced that the U.S. can do nothing. Between them, Venezuela and the Middle East alone account for around one quarter of the oil we consume. And what would Mexico do? Angry over our immigration policy, would it play the oil blackmail card?

The World in Transition

These two scenarios only consider the economic, political, and social threat. They don't consider the ecological threat. If we reach a tipping point in any combination of the crisis areas I've mentioned – political, social, economic, and ecological – the situation could quickly accelerate out of control, leaving us in a race for our lives.

The twentieth-first century will, in one way or another, be known as an age of transition that began with the events of September 11, 2001. Whether we transition into a dark age of mass murder and societal disintegration on a grand scale or whether we take civilization to a new level of social and environmental enlightenment will depend to a significant extent on the choices we make regarding energy. There will always be critics who claim that changes in energy consumption were unnecessary if they succeed. We will have to deal with far more than criticism if preventative measures aren't taken and disaster ensues. But exactly what can we do? Much of the answer lies in technology. Since the 1970s we have been steadily acquiring the wherewithal to transform our energy industry, end the Age of Hydrocarbons, and change the course of history. What it will take is the subject of the rest of this book.

PART TWO

America At the Crossroads

The personal commitment of a man to his skill, the intellectual commitment and the emotional commitment working together as one, has made the Ascent of Man.
                      Jacob Bronowski, *The Ascent of Man, 1973*

"You do whatever you can about the misery in front of you. Add your light to the sum of light."
                      Billy Kwan in *The Year of Living Dangerously*,1982

11

Alternative Technologies

To figure out how to solve a problem, after the dimensions of the problem are understood, it is often necessary to assess not only what is available to solve it, but what has to be overcome or devised for a solution to be possible. Applying these ideas to the energy trap we are in, means assessing what alternative energy technologies could provide a means of escape, then determining the obstacles that bar success, and finally working out a solution. The next few chapters will assess a variety of alternative technologies to tease out their advantages, disadvantages, challenges, and future potential as part of a new energy infrastructure that can end our dependency on hydrocarbons.

Not all technologies that are available will be considered for the plan. For example, nuclear power is excluded from consideration even though some countries, like France, have invested heavily in its potential. France currently derives two thirds of its electricity from nuclear power. Unfortunately, the extreme toxicity of nuclear waste products, their longevity, their accumulation, the costs of storing the waste, the problems of storage (which people in poorer countries could end up bearing), and the potential exploitation of nuclear waste by terrorists make present nuclear power technology an unattractive alternative. Like our hydrocarbon energy infrastructure, it appears to me to be another "time bomb," especially if Caltech professor Nate Lewis was right when he came to this conclusion, based on his peer reviewed analysis of current and projected world energy consumption:

> We could go nuclear…We have about 400 nuclear power plants in the world today. …..We'd need about 10,000 fast-breeder reactors and, by the way, their commissioned lifetime is only 50 years. That means that after we choose this route, we're building one of them every other day, or more rapidly, forever…."[1]

Fusion is also excluded from consideration, but for another reason. Barring an early breakthrough, it is not likely to be a viable source of energy for at least fifty years, probably longer.

12

Building Blocks and Stepping Stones in Transportation

The Electric Car

Among the most important technologies we could adopt to cut our oil imports and reduce pollution is the electric car. There are several reasons for reconsidering this old technology that pre-dates Otto's four-stroke internal combustion engine (ICE). Most importantly, the electric car requires no gasoline or diesel fuel to run and doesn't create air pollution. It does require electricity, much of which today is generated using coal, but electricity has distinctive advantages, as I'll explain later.

A lesser known reason to favor the electric car is its greater efficiency. Its electric motor directly produces rotary (spin) motion, the ideal mechanical motion for turning the wheels to make a car move. In contrast, the ICE used by conventional cars must convert energy from chemical to thermal to mechanical before the engine can turn the wheels to move a car, as shown in Figure 9.

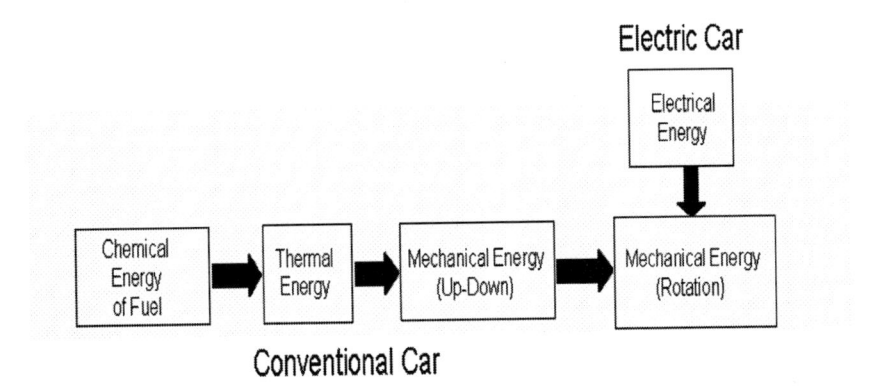

Figure 9. Energy Conversion Efficiency[1]

The electric motor is also more efficient because its engineering is much simpler. Rather than the complicated synchronization of piston and crankshaft, it uses a simple, cardinal principle of magnetism. Given that you have two magnets, you can position the magnets so that their poles – one at each end of a magnet – will repel or attract one another. Similar magnetic poles (South-South or North-North poles) will repel; opposite poles (North-South) will attract. Electric motors use either DC or AC current to set up conditions that produce spin using this principle of attraction and repulsion. Figure 10 demonstrates the basic idea, using a simple DC electric motor.

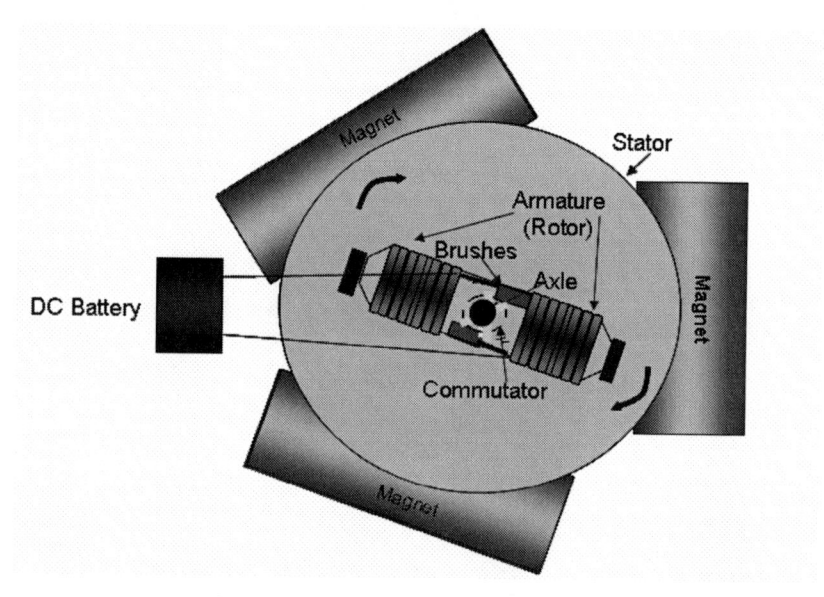

Figure 10: DC Electric Motor

The functioning of this electric motor comes down to the electromagnetic attraction and repulsion between stationary magnets and the motor's moveable armature, which contains electromagnets.

The interaction between the armature and stationary magnets is controlled through a DC battery, some stationary brushes, and a commutator, which is somewhat like a rotating switch. The DC battery activates the armature's electromagnets by supplying current to the copper wire wound round the armature's ends. The stationary brushes transfer the current from the DC battery to the armature when they contact the commutator that is rotating with the armature. The commutator, through contact with the brushes, controls both the current supplied to the armature and the armature's polarity.

Each time a magnetized end of the armature encounters a pole of a stationary magnet, the commutator switches the armature's north-south polarity causing the armature to repel from the pole it is passing and move toward the pole of the magnet that the armature will encounter next. The push and pull of the stationary magnets stimulated by the armature's reversals of polarity causes the armature to spin. The three magnets strategically placed on the stationary (stator) plate cause uninterrupted spin, once the spin is initiated.

I used a simple DC electric motor as an example. Electric motors that use alternating current (AC) create spin by taking advantage of the periodic alternations of current to reverse the armature's polarity.

Since electric cars use an electric motor, many of the auxiliary systems required by an ICE are not needed. In an electric car there is no fuel tank or fuel lines, no complex coolant system, no catalytic converter, no muffler to mute the raspy sound of exhaust, and no tailpipe. On the other hand, electric cars do add a few components besides the electric motor, mainly a power control unit and a 300v battery pack. The basic components are illustrated in Figure 11.

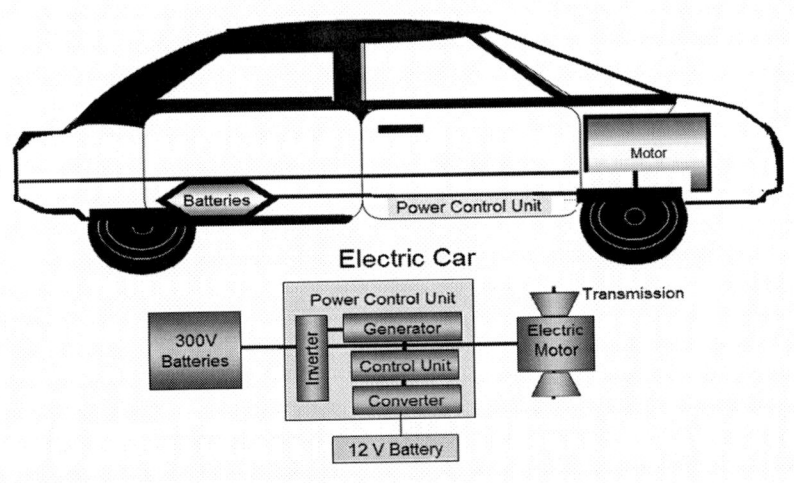

Figure 11. An Example of an Electric Car – A Basic Configuration [2]

The **power control unit** (PCU) converts DC electricity obtained from the 300-volt rechargeable battery pack to three-phase AC electricity and delivers the AC electricity to the electric motor. The PCU itself runs off electricity from a separate 12-volt battery. The current that the PCU delivers to the motor is in pulses with frequency and duration that are adjustable. Most PCUs transmit power to the engine at over 15,000 pulses per second. Because pulses of power emitted at and above this frequency are outside the range of human hearing, the motor appears to be silent.

The amount of electricity the PCU delivers to the electric motor depends on the instructions it gets from the person who drives the car. The driver "tells" the PCU to slow down or speed up by releasing or depressing the accelerator pedal. The power delivered to an electric motor ranges from none, when the pedal is not depressed, to the maximum when the accelerator pedal is pressed to the floor. The more power provided, the faster the motor's armature will spin.

A connection between the PCU and the accelerator pedal is made through a pair of devices called potentiometers or variable resistors. The potentiometers are used to modify signals that tell the PCU how much the accelerator pedal is depressed. The PCU reads the signals from two potentiometers rather than just one as a precautionary measure. It compares the two to ensure the signals are equal. If they are not equal, the PCU will not operate.

Another significant component that an electric car adds to a car's parts list is a charging system with a **regenerative braking** capability. Electric cars use regenerative braking to recapture energy for the batteries when the car coasts or brakes. The recaptured energy is transmitted to the batteries from the motor. Regenerative braking is a technological innovation that makes the electric car more energy efficient and also reduces wear on the conventional brakes.

Despite the additions, an electric car has less than one tenth as many parts as a conventional car with its internal combustion engine. And what is removed reduces many of the poisons that make the conventional car toxic. Because it is less complicated, the electric car is also easier to maintain.

Another reason for reconsidering the electric car is the cost of electricity compared to gasoline. The cost of electricity for a trip in an electric car is much lower than the cost of gasoline for the same trip in a conventional gasoline powered car, about one quarter to one third of the cost.[3]

BUILDING BLOCKS AND STEPPING STONES IN TRANSPORTATION   115

But there can be a catch – time. Take an electric car with a rechargeable NiMH battery that has a capacity of 27.4 kWh. [4] A standard home outlet of 120 volts and 15 amps can recharge the car at a rate of 1.8 kWh per hour and would thus take 15 hours to fully recharge a car that has no charge in it. However, if a 240 volt outlet at 30 amps is used, a car owner can recharge at 7.2 kWh per hour, making it possible to fully recharge in less than 4 hours. In either case, it takes a while to recharge.

The Challenges for the Electric Car: Where's the Battery?

With the electric car's advantages, anyone unfamiliar with its history might wonder why we are taking so long to replace the internal combustion engine with the electric motor. The problem lies not with the electric motor itself but with the energy storage technology used to supply it with electricity. To get an idea of what has hindered us and where we are today, we should review what happened when automakers last attempted to reintroduce the electric car as a true rival of the gasoline car. It began in California in the early 1990s after the state challenged car makers to revive the electric car. As a prod, the state established a new policy that required a certain number of cars sold here to be nonpolluting. Accepting the challenge, General Motors began developing a new electric car it called the EV1.

Recognizing that the electric car's historical weakness was its battery, General Motors gained a controlling interest in a promising new battery created by Ovonics, a company that is now famous in the alternative energy community. Though more costly, the new battery that Ovonics had created, called the nickel-metal hydride (NiMH) battery, appeared to be a significant improvement over the standard lead-acid battery used in early electric cars and used in conventional gasoline cars for almost 100 years.

What happened next is hotly disputed so I'll establish the undisputed facts first. After producing a number of electric cars, GM got the State of California to rescind its legislation on nonpolluting cars, and GM abandoned the electric car. GM would not allow any further sale of its electric cars and recalled them from its dealers. All but of few were destroyed. Other car manufacturers that had ventured into the field also abandoned their projects. General Motors' controlling interest in Ovonics batteries was taken over by an oil company, Chevron, which gained control of the patents.

While both sides of the dispute may agree up to this point on what happened, the interpretations of why GM acted as it did are different. Enraged advocates of the electric car have charged that the EV1 would have been commercially viable if GM had continued its development. They were angry that GM chose to destroy electric cars rather than allow people to buy the cars. And they have asserted that Chevron, fearing competition, used its control of the nickel-metal hydride battery patents to inhibit further development of the electric car.

In his response to these accusations, former GM employee Paul Ciszek argued that the electric car had too many battery problems. He cited several basic impediments that made GM abandon the car. They remain major challenges for the car battery industry today and have been challenges for a century:

Cost
Limited travel range between charges
Energy density and specific energy
Durability / longevity
Charging time
Safety

Each is self explanatory except for energy density and specific energy. Energy density is the amount of energy stored per volume. The greater the energy density the more power the battery can provide to operate the vehicle. Specific energy relates to the energy stored per given unit of mass. The greater the specific energy the lighter the battery can be.

Ciszek said that the new batteries cost 4-5 times more than traditional lead-acid batteries, which adversely impacted the cost of the cars. Along with the cost, Ciszek complained that the new batteries did not live up to their promise. He alleged that they were supposed to have a range between recharges of 130 miles, but that none of the prototype cars came close to achieving that mileage under normal driving conditions. Worse, while GM was developing and testing the EV1, the NiMH batteries began to fail at an alarming rate according to Ciszek. They lasted only about 6 months with regular use. Because California was demanding a battery with a five-year warranty, that would have required GM to replace 10 during the warranty period.

Even an apparent performance advantage was illusory. Although the minimum recharge time was only about two hours, Ciszek claimed that GM used a special recharging station, not a home outlet, to get that result. To fully recharge at home took 8 hours. Home electrical systems, he explained, took significantly longer because the outlets can't handle the necessary current draw for fast recharging.

Then there was the safety issue. Because of the batteries' high voltage, high energy density, and specific energy, GM decided to lease the cars and insisted that only skilled technicians could work on them. GM was worried about lawsuits. Ciszek said that safety concerns were the reasons why GM decided to destroy nearly all of the cars rather than let anyone buy them.

As a final discouragement, Ciszek cited public apathy. Few people seemed to be interested in trying out an electric vehicle, much less owning one.

Breakthroughs that Affect the Electric Car

The end of the EV1 has had strong repercussions. Until recently, finding an electric vehicle in the United States above the size of a golf cart has been difficult so why even suggest we try again to revive the electric car? The answer is new technology and dedicated companies eager to exploit it. An example is Tesla Motors, a company named after the father of AC electricity. Its first model, the Tesla Electric Roadster, was offered for sale in 2006. Over 100 have been sold so far, though there have been delays in delivery. They go for about $100,000. If successful, Tesla Motors plans to build a lower price model for the average customer. It should cost around $30,000.

The specs on the Tesla Electric Roadster are remarkable. Tesla Motors claims its Roadster can go from 0 to 60 mph in 4 seconds, reach a top speed of 130 mph, and can go up to 250 miles before having to recharge its battery. They claim the Roadster has 248 horse power[5] and can fully recharge in 3.5 hours. What makes much of this capability possible is Tesla Motors' lithium-ion battery, which the new car company claims has a battery life in excess of 100,000 miles. Each of the battery's specifications addresses key challenges facing energy storage technology: range between charges, energy density, durability/longevity, and charging time. And it claims it can bring the cost down by its strategy, which is to offer its car initially at a premium to wealthy customers, get a foothold in the market, and then mass produce cars to reduce costs.

If all of this is true, has Tesla Motors finally made the breakthrough in energy storage technology that will allow the electric car to compete commercially with the gasoline car? It's possible, but there is more than a hint of trouble along with the promise. Tesla Motors claims that it has combined basic lithium ion battery technology with its "own battery pack design to provide multiple layers of safety."[6] Why the multiple layers of safety?

To understand the reason safety was mentioned, and to better assess the Roadster's battery in general, you need to know a bit more about how batteries operate. The job of a battery is to produce an electrochemical reaction internally that allows the battery to deliver current and voltage from its terminals when needed. Different batteries perform this job in slightly different ways.

The most used car battery, the lead-acid battery, can serve as an example. Internally, a lead-acid battery cell is composed of two plates, each connected to a terminal. One plate called an anode is made of lead; the other called a cathode is made of lead dioxide. Inside the battery, the plates are immersed in an electrolyte containing sulfuric acid (H_2SO_4). The sulfuric acid reacts with the lead on the lead anode plate to produce lead sulfate ($PbSO_4$), freeing electrons in the process and causing the anode to become negatively charged. The electrons collect on the plate's surface along with the lead sulfate. The sulfuric acid also reacts with the lead dioxide cathode plate to deplete electrons, producing lead sulfate and water in the process. The depletion of electrons causes the cathode to become positively charged.

The negatively charged electrons on the lead anode plate are drawn over the external, closed circuit to the positively charged lead dioxide cathode plate. As the battery is used up, the amount of lead sulfate on both plates builds up, and water builds up in the electrolyte while sulfuric acid is reduced. Amazingly, this process is reversible. One of the reasons that lead-acid batteries are used in cars is that they are rechargeable; that is, you can reverse this reaction and through the reversed reaction replenish the battery.

All conventional batteries, from the very first, have behaved basically the way the lead-acid battery does when it comes to delivering electricity.[7] But the different chemical properties of the battery cells introduce subtle and sometimes not so subtle differences in capability. For example, because of the lithium-ion battery's chemistry, it has a higher energy density than most other batteries. The lithium-ion battery in a Tesla Roadster has a capacity of 52 kilowatt-hours (kWh) compared to the improved NiMH battery used by the Toyota Rav4 EV, which has a capacity of 27.4 kWh. This difference is the main reason the Tesla Roadster has a much greater range between recharges than the RAV4 EV. Another way of making the comparison is through electrical storage capacity per kilogram (kg). Typically, lithium ion batteries can store 0.15 kWh of electricity per kg of battery compared to 0.10 kWh per kg for the NiMH battery and 0.025 kWh per kg for the lead acid battery.

While the lithium-ion battery gets high marks for high energy density, it raises concerns with respect to safety, in part because of its high energy density. Internally, the lithium-ion battery has an anode plate made of carbon, a cathode plate made of lithium cobalt dioxide ($LiCoO_2$), an electrolyte made from an organic solvent, and a separator made of plastic. What can make this combination a safety concern is the lithium. It can cause an explosion if the battery is heated above its temperature range. To avoid problems, lithium-ion batteries usually have built-in protection to prevent conditions that cause overheating or unacceptably high-voltage. Tesla Motors claims to have increased protection by preventing the failure of one battery cell from causing a cascade failure in other cells.

Lithium-ion batteries have several other known disadvantages. There is a limit on how deeply these batteries can be discharged.[8] The battery is not supposed to be discharged below a certain voltage to avoid irreversible damage. And it has been found to go bad after a few years, even if left unused on a shelf. The problem is reduced by keeping the battery from sitting on the shelf fully charged. But the claim by Tesla Motors of a 100,000 mile battery life is exceptional. Lithium-ion batteries are also more expensive than lead-acid batteries, because of their chemical composition.

On the other hand, the known advantages of standard lithium-ion batteries are also significant, such as the previously described high energy density, and their high specific energy which allows a car manufacturer to use a lighter battery. Lithium-ion batteries also have a recharge advantage. They can recharge relatively quickly. Yet the battery's self-discharge rate is lower than that of a NiMH or lead acid battery.

Until it is exercised long term under normal driving conditions, the durability and reliability of Tesla Motor's lithium-ion battery design and of the entire power train will remain uncertain. But there is reason to be hopeful in any case.

Batteries like the lithium-ion are improving at a breath-taking speed, thanks in part to their demand, not in the car market but in other markets such as the red-hot laptop computer market. An example of the continued improvement of the lithium-ion battery is the lithium-ion polymer battery which replaced a liquid electrolyte with a less costly solid polymer and improved the battery's resistance to physical damage.

Other promising new technology is also being tested including super capacitors, organic batteries, flywheels, vanadium flow batteries, and the fuel cell, which is covered in detail in Chapter 17. Because of the progress already made, the future for energy storage technology is looking increasingly bright.

So we should not discount the all-electric car. In a few years, it may become a major player.

The Hybrid Car

In the meantime, we don't have to wait for the ideal electric car. We have an alternative. The gasoline-electric hybrid car is already doing quite well in the competitive commercial market. It is what I call a **bridging technology**. It provides a near-term solution to some of the battery problems that have held the electric car back. For example, by providing a constant recharging capability while it is moving, the hybrid prevents its batteries from being deeply discharged, which extends battery life and improves battery performance. Overall, thanks to its combination of an internal combustion engine and an electric motor, these hybrids decrease gasoline consumption, increase efficiency, and reduce pollution.

Because it has many characteristics of a conventional car, many of the subsystems of a gasoline-electric hybrid are comfortably familiar, such as the fuel tank, on board computer system (used for energy management and system control), energy storage unit, generator, transmission, gasoline engine, thermal management (cooling) system, and emissions controller (catalytic converter).[9] Most of these systems are illustrated in Figure 12, a drawing taken from the government's Energy Efficiency and Renewable Energy program (EERE) website.

[Figure showing hybrid car components with labels: Accessories, Energy Management & System Controls, Body Chassis, Fuel Tank, Thermal Management System, Hybrid Power Unit, Traction Motor, Energy Storage Unit]

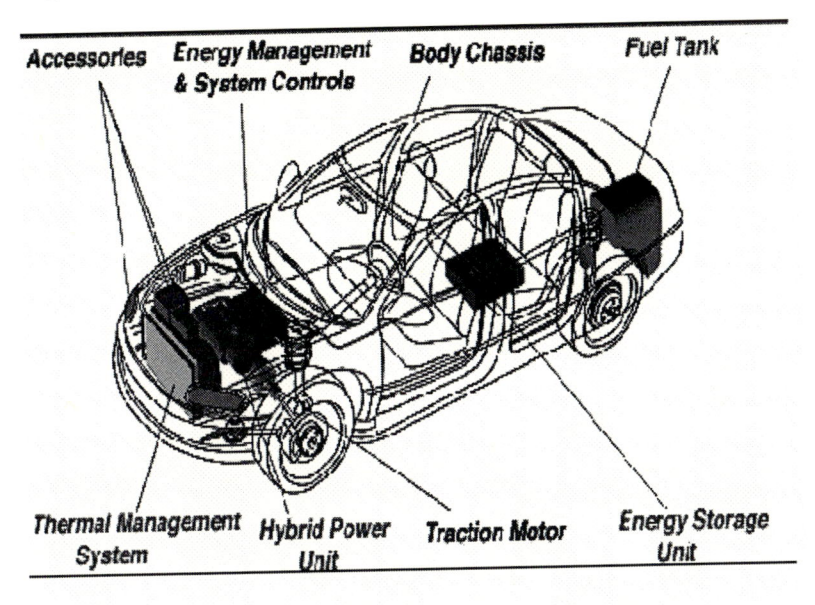

Figure 12. Components of a Gasoline-Electric Hybrid Car[10]

What's unique and novel to most people are the hybrid power unit additions. The components in Figure 12 are configured and extended in the two basic hybrid electric vehicle (HEV) designs: the **parallel HEV** and the **series HEV**.[11]

Hybrid Electric Vehicle (HEV) Designs

A good starting point for understanding and assessing these two HEV designs is the conventional power train. The power train starts at the **engine**. Power is transferred from the internal combustion engine through the **clutch**, the **transmission**, the **drive shaft**, and the **differential** to the wheels.

The clutch is used to connect the engine and to disengage it from the transmission. The engine is disengaged when the car is idling. The transmission translates the engine's limited range of rotational speeds into a broader range of speeds with which to turn the drive shaft. The differential takes the rotational speed of the drive shaft and refines it to turn the wheels. The engine, clutch, transmission, drive shaft, and differential as shown in Figure 13 form the power train.

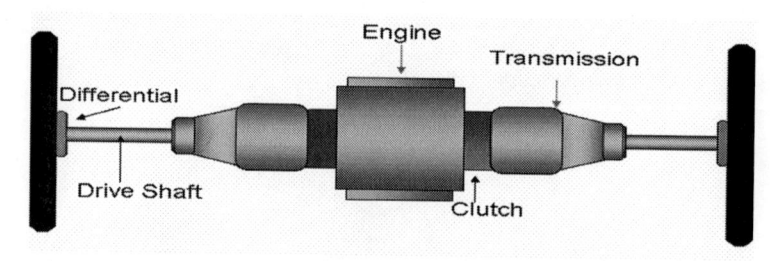

Figure 13. Power Train [12]

In the **parallel HEV** both the electric motor and internal combustion engine (ICE), though working together as a hybrid power unit, are independently connected to the transmission. In contrast, the **series HEV** uses only the electric motor to transmit its power to the transmission. The ICE simply assists the electric motor by turning a generator that charges the batteries on which the electric motor depends for electricity. In operation, the series HEV acts more like an electric vehicle than the parallel HEV. The downside for series HEVs is that they require larger, heavier battery packs than parallel vehicles do, which is why they are rarer. Besides the parallel and series HEVs, a design that combines series and parallel attributes has also become commercially available and very successful.[13]

Hybrid Tricks that Save Gas and Reduce Pollution

The hybrid car's greater fuel efficiency and cleaner emissions are not just the result of using an electric motor. Another contributing feature is its smaller internal combustion engine (ICE), compared to most conventional cars. It might be supposed that the smaller ICE would cause the hybrids to be considerably slower to accelerate and less able to handle steep hills. That is usually not the case. The hybrid is able to keep up with many vehicles that have larger ICEs because it can draw on the extra power it needs from its electric motor, as long as its battery has charge to power it. It is able to get away with this arrangement because most drivers use the peak power of their engines less than one percent of the time.

Besides using a smaller, more efficient engine, manufacturers of parallel, and combined parallel/series hybrids in particular, use a number of other tricks that make their hybrids more fuel efficient and less polluting. Some of the tricks would help any car. Others are only possible because of the hybrid electric motor and battery design. The most noticeable, perhaps, is the first in my list of hybrid tricks, the hybrid's ability to turn off its engine instead of idling.

1. Energy Source Switching and Shutoff – Because the hybrid car's computer has the luxury of selecting between the gasoline engine and the electric motor or using both, these hybrids can shut off the gasoline engine when the car is stopped, move the car forward under electric power only, and then turn the engine back on again as needed. This saves a lot of gasoline, especially when the car is waiting at a stop signal or in stop-and-go traffic.

2. Catalytic Converter Stays Hot – For decades, starting a gasoline car produced a larger than normal amount of pollution because the catalytic converter was too cold to begin catalyzing combustion products. This could have been an even greater problem for the hybrid because its gasoline engine is also turned off when it isn't needed, as for example at a stop light or while coasting. But fortunately, a solution to the problem was found.

The solution is to use variable conductance insulation and phase-change heat storage material patented by the National Renewable Energy Laboratory (NREL). The material keeps the catalytic converter hot for more than 17 hours, yet allows heat to flow during peak engine loads to prevent the catalytic converter from overheating. This solution allows 95% of all auto trips to begin with a sufficiently hot catalytic converter. For conventional cars, a hot catalytic converter means fewer emissions when the engine is started. For hybrids it means in addition that the catalytic converter remained hot enough to work efficiently despite the gasoline engine being periodically shut off during a trip.

3. Coordinated Use of Electric Motor and Gasoline Engine – The hybrid can coordinate, shut off, and switch its power sources for greater fuel efficiency when idling, at higher speeds, during acceleration, and on hills.

For example, at higher speeds when the gasoline engine is most efficient, it takes over as primary; whenever the car slows down, the gasoline motor will shut off instantaneously; and when going uphill, both the electric motor and gasoline engine are usually fully engaged, unless the uphill grade is so long that the electric motor will disengage because the charge in the battery has been used up.

4. **Regenerative Braking** – Hybrids are able to generate electricity every time they slow down or brake. In fact, the electric generator helps hybrids slow down when they brake which reduces wear on the brakes.

5. **Aerodynamic Design** – Much of the work of the engine goes into moving the car forward against air resistance. The faster a car moves the more air resistance it encounters. This is clear to anyone who sticks a hand out the window of a car on a freeway. The impeding force from the air resistance is called aerodynamic drag or **air drag**. Reductions in air drag reduce the amount of work the engine must perform overcoming this impeding force and consequently the amount of gasoline used is reduced.

Although not all hybrids have better aerodynamics than other cars, most use at least some tricks to reduce air drag. Narrowing the car's front end and using curvature in the body design reduce air drag. The effect is somewhat like the reduced air drag of a jet plane or rocket with a tapered nose. Another way to reduce air drag is with air dam spoilers, which are aerodynamic devices that change the direction of air in order to reduce airflow and therefore drag beneath the car. Also used are special high inflation tires that allow the wheels to roll more easily.

Hybrids also use advanced technology found in some conventional cars to improve mpg and performance including variable value timing and electronic control (VTEC), dual and sequential ignition (DSI), continuously variable transmission (CVT), and light weight advanced materials.

The Toyota Prius

The original hybrids produced by the Japanese, the Honda Insight, Honda Civic Hybrid, and Toyota Prius, established their dominance of the hybrid market. All three are still excellent examples of hybrid technology.

BUILDING BLOCKS AND STEPPING STONES IN TRANSPORTATION   125

I'm going to highlight the Prius because it reflects the basic hybrid technology and has some unique features that contribute to the technology.

Like the Civic Hybrid and Insight, the Prius meets California's super ultra low emissions vehicle (SULEV) standard and the federal government's advanced technology partial zero emission vehicle (AT-PZEV) standard. Meeting both these standards means that its pollution emissions are 90% less than that from a conventional gasoline car.

Besides meeting these low emissions standards, the Prius delivers significant gasoline mileage economy. The Prius' EPA figure for fuel efficiency was originally 60 mpg city and 51 mpg highway.[14] However the EPA improved its method of measuring mpg and revised its estimate of Prius fuel efficiency to 47 mpg overall, which is more in line with what many owners report.

Nearly everything about the Prius makes it an efficient car.[15] To start with, thanks to its shape and surface exterior, the Prius has a low drag coefficient of 0.26. The lower the drag coefficient the less air resistance the car encounters and the more fuel efficiency it has. By comparison, the average midsize passenger car has a drag coefficient of 0.30-0.35. And SUVs have a drag coefficient of 0.35-0.45.[16]

The car's electronic displays, which are in the middle of the dashboard, promote more efficient driving

Figure 14. Multi-Information Display[17]

They tell the driver how many miles per gallon the car is getting currently and at previous points in time. With this feedback that connects road conditions and driving habits with mpg, a driver can figure out what driving habits increase fuel efficiency. This dashboard display is also a good introduction to how Toyota's Synergy Drive operates.

Prius's Planetary Transmission

Of all the components that comprise the Prius, the one that stands out from an engineering point of view is the transmission. The transmission is the heart of Toyota's **Hybrid Synergy Drive**. It allows the Prius to operate both as a parallel and series hybrid. The key to its versatility is Toyota's brilliant adaptation of the conventional transmission planetary gear system so that it becomes a power-split transmission, which can accept power from either the electric motor, the gasoline engine, or both.

The conventional transmission planetary gear set consists of four main components: a **sun gear** (at the center), a **planetary carrier, planetary or pinion gears** (which are connected to the planetary carrier and arranged around the sun gear), and a **ring gear**. Various interactions of two or more gears produce the gear ratios used to control the speed of the drive shaft. Toyota took this concept in a new direction, as illustrated in Figure 15.

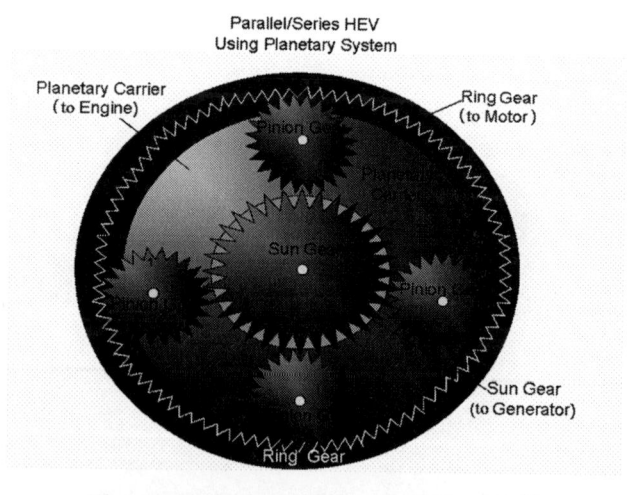

Figure 15. Toyota's Planetary System[18]

Its engineers connected the shaft of the planetary carrier to the engine, the shaft of the ring gear to the motor, and the shaft of the sun gear to the generator. And they connected both the motor and engine through the ring gear to the drive shaft and differential.

Because the motor, generator, and engine can engage and disengage one another through the gears and planetary carrier, it is possible to create different behaviors to fit different driving situations. Just before the car begins to accelerate from a start with the engine off, for example, the electric motor engages the ring gear so that it starts spinning with the motor. The ring gear engages the pinion gears starting them spinning. The pinion gears engage the sun gear starting it and the generator spinning. Meanwhile, the planet carrier, connected to the engine, is still stationary because the engine is not running. With acceleration, the sun gear suddenly changes speed and causes the planetary carrier to be engaged to turn and start the engine. If the car accelerates quickly, the motor will draw extra power from the batteries. It is this revolutionary use of gears, which connects the engine to the generator and to the transmission, that allows the Prius to operate both as a parallel hybrid and a series hybrid. With this scope, the Prius can optimize the differing strengths of parallel and series hybrid technology to improve fuel efficiency to an extent that is so far unrivaled.[19]

In addition, under the hood the Prius has a small but highly efficient 76 horsepower 1.5 liter four cylinder engine, an electric motor, a generator, a nickel-metal hydride (NiMH) battery (before 2008), a power control unit, regenerative braking, and a continuously variable transmission. The system generates 110 horsepower, enough power for the car to accelerate from 0 to 60 mph in 10 seconds and go 105 MPH.

Challenges for Hybrid Cars

Although they have a lot to offer, hybrids like other alternative vehicles have faced significant challenges in competing with well-established conventional cars. To a lesser extent, they still face the challenges associated with energy storage technology: cost, energy density, range between charges, durability/longevity, charging time, and safety. But they also face a challenge that ironically grows out of their successful advertising campaign.

To succeed in selling hybrids, the car manufacturers concluded they would have to distance their cars from the distasteful image of the electric car as "inconvenient." A major obstacle to selling electric cars was the reluctance of potential buyers to purchase a car that had to be constantly plugged-in to electrical outlets. Potential customers expressed the fear that they won't be able to find an outlet when needed. Unfortunately, by boasting that their hybrids don't need to be plugged in, the hybrid car manufacturers may be inhibiting a further breakthrough in fuel efficiency.

Potential Breakthroughs

Tinkerers who have altered the Prius so that it has both more battery capacity and the capability to be plugged into an electrical outlet claim that the Prius plug-in can go 100 miles on a gallon of gasoline. That increase, along with a decrease in the time required to recharge the batteries, should remove much of the inconvenience of plugging the cars in. It should be more convenient than stopping at a service station, given the superior gasoline mileage. Other near future changes that improve efficiency are also possible including greater energy recapture from the internal combustion engine and the use of a solar blanket as an additional source of electrical power.

The Future of Gasoline/Electric Hybrid Cars

Recognizing their importance, the federal government has encouraged hybrid purchases by offering a tax credit to buyers of hybrids that get over 40 mpg, although the offer was limited to a certain number. In addition, several states have joined in providing tax breaks and other incentives such as allowing these hybrids into the carpool lanes when the car is only occupied by the driver. 170,000 Prius hybrids were sold in the U.S. in 2007 alone, despite the loss of tax incentives for Priuses bought in the latter part of the year after the quota was reached. Although this is a small fraction of the new cars and trucks sold each year in the U.S., the number sold has been enough of an incentive to encourage other car manufacturers to increase their involvement, especially since the growth in sales has been accelerating not only in the U.S but worldwide.[20] In 2005 Daimler–Chrysler, Ford, and General Motors joined Honda and Toyota in offering hybrids. And more are on the way.

Gasoline/electric hybrids represent an ingenious way to make the relatively inefficient internal combustion engine (ICE) not only more efficient but less polluting. This comes at a time when improvements in fuel efficiency in the conventional ICE car seem to be getting smaller and more difficult to realize. There is little doubt that the hybrid or electric car can be instrumental in significantly reducing imported oil and pollution. But unfortunately, the U.S. automakers that make hybrids are making mainly hybrid SUVs, which get far fewer miles per gallon than cars like the Prius. Furthermore, American car companies appear more interested in a very different bridging technology, which we'll examine next.

Diesels, Alternative, and Flexible Fuel Vehicles

The alternatives favored by American car companies are called flexible fuel vehicles (FFV). They need to be distinguished from the alternative fuel vehicle (AFV). Sometimes the two terms FFV and AFV are used interchangeably but there is a difference between them. AFVs use a fuel other than gasoline or conventional diesel, while FFVs use more than one type of fuel at a time, one of which is usually gasoline. The most prevalent AFVs use biodiesel, propane,[21] or compressed natural gas (CNG). Biodiesel, propane, and CNG are all cleaner fuels than gasoline, but they are hydrocarbons usually obtained from non-renewable sources. Claims vary on how much of a reduction in pollution you can expect from these alternatives. Compared to gasoline, CNG may reduce tailpipe emissions of nitrogen oxides (NOx) by 20-25%, carbon monoxide (CO) by 50-90%, other ozone-causing pollutants by 50-75%, and particulate emissions by almost 100%. Tested tailpipe emissions from propane-powered vehicles suggest it can cut emission of NOx by 20-50%, emission of CO by 45%, and other ozone-causing pollutants by about 60%, when compared with the average gasoline vehicle.[22]

While propane and CNG vehicles still have conventional ICEs, diesel cars use diesel ICEs. Diesel engines have long been associated with heavy trucks, unpleasant odors, and high particulate counts. But unlike models sold in this country during the 1970s and 1980s, modern passenger cars with diesel engines are quiet, smooth, more responsive than older models, almost entirely free of diesel odor, more fuel efficient, and considerably cleaner, all thanks to their use of turbocharged-direct-injection (TDI) computer-controlled systems.

With TDI, fuel is directly injected into the combustion chamber. Advanced fuel injectors atomize the fuel into a fine mist. The combustion chamber swirls the air and fuel mixture, homogenizing it further. The computerized electronic management system controls the entire engine operation, including the turbocharger, and fine-tunes the process to ensure fuel efficiency and emission control. The TDI increases fuel economy in diesels by up to 20% over conventional diesels while making the ride seem more like that of a car with a gasoline engine.[23] Making them even less conventional, diesel engines don't need to run on diesel at all. In fact, the original diesel engine, which was invented circa 1893, actually ran on peanut or vegetable oils. Diesel engines retain that capability today with minor modifications. Unfortunately, since vehicles with diesel engines comprise only about 1% of the U.S. passenger car market[24], service stations in the U.S. that sell diesel fuel are more difficult to find.

The other type of vehicle, the flexible fuel vehicle (FFV) comes in two varieties: **bi-tank** and **blended fuel models**. Bi-tank models, as their name implies, store their fuels separately, which make them a type of hybrid. The most typical example is the natural gas-gasoline FFV. The car is an FFV hybrid because natural gas requires a pressurized tank separate from the gasoline fuel tank. Blended fuel models, in contrast, require only a single tank since the fuel is blended before it enters the car. They don't need to store their fuels separately. Ethanol or methanol and gasoline are the leading blended fuels used. Of the two FFV types, the blended fuel model is the one favored by American car manufacturers because they require only minor modifications of existing car designs. For the most common blend, ethanol with gasoline, no modifications to conventional cars are necessary so long as the blends do not contain above 10% ethanol (E10). Even above 10%, fairly modest modifications are required, such as a sensor that can analyze the fuel composition and adjust the fuel injection during ignition. And the leading alternative fuels, ethanol and methanol, can easily be blended with gasoline in variable quantities. Blends of 10, 15, and 85% are the most prevalent. In Brazil even 100% ethanol is found, which blurs the distinction between an FFV and an AFV.

Challenges Facing FFV and AFV: Fuel

So should we prefer FFVs and AFVs to hybrids as a means of reducing oil imports? There are problems. FFVs that use over 85% ethanol in colder climates don't start easily. Conventional cars converted to an alternative fuel in the U.S. still contain at least 15% gasoline in order to have enough power for cold starts. Neither FFVs nor AFVs are as efficient in reducing pollution as the Japanese gasoline-electric hybrids. And both the FFVs and AFVs have other drawbacks given their fuel sources. Propane, CNG, ethanol, and methanol are hydrocarbon fuels currently obtained from nonrenewable sources. And though FFVs and AFVs can potentially run on cleaner, renewable fuels, that capability depends on the state of the renewable biofuel industry and whether it can produce biofuel in sufficient quantities.

13

Biofuels and Coal: Catalysts of Transition?

So what is the state of the renewable fuel industry? It can be characterized as immature but promising. Until recently, renewable fuel manufacturing in the U.S. didn't receive much press attention, despite its potential importance. Renewable fuels have the capability to directly replace oil, coal, and gasoline without requiring a costly, new energy infrastructure. They can be bridging fuels as well. That is, they can be converted into hydrogen, which some argue is the fuel of the future. They can also be made by combining hydrogen and carbon dioxide, using a process that will be described later. Currently, there are three leading methods for obtaining renewable fuel:

1) Thermo-Chemical Gasification. General plant matter is the input. The plant matter is heated, but not burned, to break down its structure into various gases, liquids, and solids. These products are manipulated using special chemicals and catalysts to produce liquid fuels and gases that range from methane to alcohol.

2) Biochemical Bio-Digestion. The bio-digestion pyrolytic method uses inputs ranging from corn and sugar cane to sewage, depending on the exact process and the digesters. In fermentation (one example of a biochemical process), bacteria, yeasts, and enzymes are used to break down carbohydrates in corn or sugar cane to produce ethanol. Other biochemical processes use bacteria to break down sewage and landfill biomass into methane.

3) Chemical Gasification. In this chemical process, biomass oils such as soybean and canola oil are chemically converted into a liquid fuel similar to diesel fuel and into gasoline additives, using heat and catalysts. Waste cooking oil from restaurants is an example of an input source that can be used to make biodiesel. The renewable biomass used in these processes can range from corn and sugar cane to sewage, manure, organic garbage from landfills, wood, and crop residues.

Breakthroughs Needed in Renewable Fuel Production

There are three major challenges that face the renewable fuel industry for which breakthroughs are needed.

    Energy Production Ratio (EPR)
    Mass Production / Scalability
    Cost

What makes the renewable fuel industry immature is the Energy Production Ratio (EPR) also known as the Energy Payback Ratio and Energy Profit Ratio.[1] The EPR is the ratio of energy output to input. To date, the three leading methods of producing biofuel from biomass require too much energy. The ratio of energy output to input is not good enough. For example, the EPR for producing ethanol in the U.S is from 1.25:1 to 1.8:1. That is, we can expect to get 25-80% more energy output than is required as input.[2] That may sound impressive, but it isn't. It is quite low compared with the EPR for oil, which is between 10:1 and 15:1, depending on the quality of the oil. Admittedly, the calculation of the EPR for ethanol (as well as for other biofuels) remains somewhat controversial, but the low EPR calculation seems to be borne out in the U.S. and may be even worse than suggested here.[3] On the other hand, as this technology receives greater attention, improvements in EPR may come. The other major breakthroughs needed to make renewable fuels a viable option are the production of renewable fuels in the quantities that would be needed at a cost that the average consumer can afford to pay. These remain huge hurdles.

Synthetic Natural Gas (Syngas)

Some fairly recent announcements are raising hopes that we could create a workable renewable fuel program here in the U.S. through new technology, but there are caveats. One announcement made in May of 2005 by Fairchild International Corporation, the holding company for SynGas of Canada, suggested a dramatic breakthrough. Fairchild announced a proprietary technology that could produce synthetic natural gas (chiefly methane) at a cost below that required to refine natural gas. Being proprietary, they disclosed little except their claims.

Not only did Fairchild claim it could produce syngas at a lower cost than other companies can refine natural gas, it claimed it could build a factory to produce syngas at a cost that would be less than that required to build a new natural gas refinery.[4] Since natural gas, like oil, is a limited resource, this could be a very significant announcement. It also suggests that Fairchild (which changed its name to Syngas International in 2006) had found a way to raise the EPR for biomass conversion to 10:1, which is the EPR for natural gas, according to some calculations.[5] It remains to be seen. Determining what claims are genuine and what are overblown are among the difficulties in making assessments, especially when supporting information is kept proprietary.

Biofuel from Microbial Fuel Cells

More immediately credible to date – because of the publication of papers in reputable journals – is an announcement made by University of Pennsylvania researchers in the summer of 2004 that they had developed a microbial fuel cell (MFC) that could produce fuels like ethanol and hydrogen. The MFC holds the promise of producing beneficial supplies of energy from smelly toxic bio-hazardous waste often wantonly dumped into sewage and toxic "lagoons." The microbial fuel cell is a variant of the biochemical bio-digestion method for producing biofuels. Its basic purpose, however, was not to produce ethanol and hydrogen but to generate electricity.

A microbial fuel cell is comprised of two chambers: the anaerobic anode chamber and the aerobic cathode chamber. In the anaerobic chamber, anaerobic bacteria of the family geobacteraceae are provided with organic matter called substrate, consisting mostly of wastewater and glucose. The bacteria oxidize the substrate during consumption, producing biofuel gases and hydrogen.[6] But that is just the start of the process. During oxidation, the anaerobic bacteria also transfer electrons to the anode. By adding an external circuit between the anaerobic anode chamber and the aerobic cathode chamber, researchers were able to produce a current with these electrons. Meanwhile, protons produced in the anaerobic chamber during oxidation diffuse to the aerobic chamber through a proton-permeable NafionTM membrane. In the aerobic cathode chamber, the electrons and protons combine with oxygen to form water.

The microbial fuel cells described in this experiment encountered some setbacks. In particular, dissolved oxygen leaked from the aerobic chamber into the anaerobic chamber through the membrane. The presence of oxygen from the aerobic chamber can inhibit the growth of anaerobic bacteria. To create a more oxygen free environment for the anaerobic bacteria, MFCs were kept in a temperature-controlled room at around 30°C (86°F), and an oxygen scavenger called cysteine was added to the chamber. Cysteine reacts with oxygen to form a disulfide dimer, an action that has the effect of taking free oxygen out of the liquid environment.

While the oxygen scavenger helped to reduce the effects of oxygen contamination from the aerobic chamber, it did not completely eliminate the effects of oxygen leakage into the anaerobic chamber. Furthermore, the reported production of electricity using the prototype MFC systems was relatively small – between 10 and 50 milliwatts per square meter (mW/m²) of electrode surface.[7]

Nonetheless, even as originally conceived, the process has considerable potential, especially because it has multiple uses as a provider of biofuel and electricity and as a cleaner of wastewater. It was reported that the bacteria could remove up to 78 per cent of the organic waste matter present in wastewater (such as leftover rotting food, sewage, and bio-hazardous materials). Each year in the United States, about 33 billion gallons of domestic wastewater is treated at a cost of $25 billion. Much of the cost is for the energy to run the treatment plants, so the MFC could save a lot of money.[8]

The announcement of progress on the MFC was exciting enough, but the MFC was about to get even better. In a second ground-breaking paper, the University of Pennsylvania researchers announced a surprising find; they discovered a method of boosting the hydrogen output of the MFC to what appears to be an EPR of 5:1. More of this discovery will be described later in the chapter on hydrogen production.

Ethanol Production in Brazil

Other innovations are coming from south of the border. Today Brazil has the distinction of being a world leader in the production of ethanol as fuel for automobiles, thanks to policies that date back to the 1970s. Ethanol is offered in most Brazilian service stations. As a snapshot, in

August 2006 pure ethanol (E100) was selling for 53 cents per liter (about $2 per gallon), compared with 99 cents per liter for gasoline (about $3.75 per gallon).[9] It is estimated that Brazil saved 55 billion dollars (U.S.) on oil imports from 1975 to 2003 as the result of its ethanol program.[10]

For their ethanol production, the use of sugar cane rather than corn seems to be one reason Brazil has been able to do so well. From the 1970s to the late 1990s, ethanol yields rose from 242 to 593 gallons per acre,[11] thanks to genetic manipulation and smart cultivation practices. As a consequence, today ethanol production in Brazil is apparently profitable, with Brazilians eager to sell ethanol to the world. Adding to the good news for Brazil, thanks to flexible fuel vehicles, car owners have a choice. They aren't locked into 100% ethanol or 100% gasoline. They have the flexibility to select what best fits their pocketbooks.

So why don't we emulate Brazil? There are several reasons. Aside from the Hawaiian Islands and some Southern states, sugar cane is difficult, if not impossible, to grow in the U.S. There are also down sides to using ethanol from the environmental viewpoint. Ethanol can produce smog.

What is more worrisome to many is that if ethanol is derived from agricultural crops, such as sugarcane and corn, it will force farmers to choose between using their land to produce food crops and to produce fuel. For environmentalists in Brazil, expansion of ethanol production is also worrisome because it is putting pressure on the rain forest.

Then there is the question of efficiency. Brazil produces about 400 million metric tons of sugar per year, mostly for ethanol. One metric ton of sugar cane is used in the production of 79.5 liters of ethanol.[12] That may not be the most efficient use of land.

Finally, it would take far more ethanol than can conceivably be produced to fuel all the conventional cars on the road today. Given the drawbacks of ethanol, the emphasis that American automobile manufacturers put on FFVs seems misplaced. However, if the EPR, cost, and scale of production can be improved without dangerously taxing agriculture and nature's biomass, ethanol should be considered. It could be included in a near-term, multi-solution strategy to reduce oil imports and pollution.

By 2007 the production of renewable fuels, particularly of ethanol, began to increase noticeably. Its production is figuring more prominently in the government strategy to reduce our import dependency.

Using Coal Gasification and Fracturing to Obtain Oil

Our government has also been looking at ways to increase domestic production of oil. A possible solution gaining favor is coal-to-oil conversion.

The most common process used for the production of oil from coal came out of Germany during World War II, born of German desperation. Once Germany lost its foothold in the Middle East, its need for oil imperiled its war effort. What it still had to draw on was coal. The process German scientists pioneered to obtain oil from coal is called the Fischer-Tropsch process. It is a coal gasification technique that uses steam and oxygen passed over coke at high temperatures and pressures to produce hydrogen and carbon monoxide. These gases are then assembled into natural gas or oil.[13]

Another competing process that may prove promising is called Low Temperature Carbonization (LTC). It was developed in the 1920s by Lewis C. Karrick at the U.S. Bureau of Mines. Simpler than the Fischer-Tropsch process, the Karrick LTC process heats coal to 650-1380°F in the absence of oxygen and then distills oil from it. It's been suggested that LTC in conjunction with the Fischer-Tropsch process could produce about a barrel of oil per ton of coal along with 300 BTU of natural gas and 1500 lbs of solid smokeless char.[14] Solid smokeless char can be substituted for coal in utility boilers and substituted for coke in steel smelters.

Another way to increase domestic oil production is through extraction of more oil from oil shale and mud stone. Perhaps the least expensive extraction process to date is fracturing, which involves injecting water and sand into the rock layer to make it more porous so the trapped oil can be bled out.

The problems with using coal to produce oil or with using fracturing to extract harder-to-get domestic oil are cost, pollution, and the possibility of delaying a transition to alternative energy sources in the transportation and stationary power sectors. Moreover, obtaining oil from coal is not the most efficient use of coal. Nonetheless, coal-to-oil conversion is enticing as a short-term option because of coal's abundance in the U.S.

14

Broadening the Stationary Power Base

Like the transportation sector, the stationary power sector has alternatives to fossil hydrocarbons. Among the most environmentally friendly are energy conversion devices which use wind, solar, geothermal energy, or ocean waves to generate electricity. Several of these devices are mature enough to be commercially in use or are part of demonstration projects. They include wind turbines, solar cells, thermocline energy systems, and hybrid systems that combine conventional gas turbines with alternative technology. Perhaps the oldest of the alternatives to exploit natural energy sources is wind technology.

Wind Turbines

In centuries past, the energy from wind and flowing rivers was used to turn windmills and water wheels that supplied the mechanical power for milling wheat and other grains. In the 19th century the same basic idea was first used to produce electricity from hydroelectric generators and wind turbines.[1]

Modern wind turbines are comprised of three basic components: rotor blades, mechanical-electric conversion equipment, and a tower. **Rotor blades** convert the wind's energy into rotational energy with a specific axis orientation. The most popular design uses a horizontal axis of rotation like that of a traditional windmill. The **mechanical-electric conversion equipment**, which connects to the rotor blades, transforms mechanical energy into electrical energy. This equipment includes a gearbox, a brake, a yaw drive, a generator, a controller, and support equipment, all of which is enclosed in a nacelle that protects it. The **tower** is used to lift the wind turbine to a height where the wind is stronger and often steadier than at ground level. The tower can be fixed or tilted up. Tilt ups are often preferred today because they can be easily taken down for maintenance on the ground.[2] All of these components and relationship to one another are shown in Figure 16.

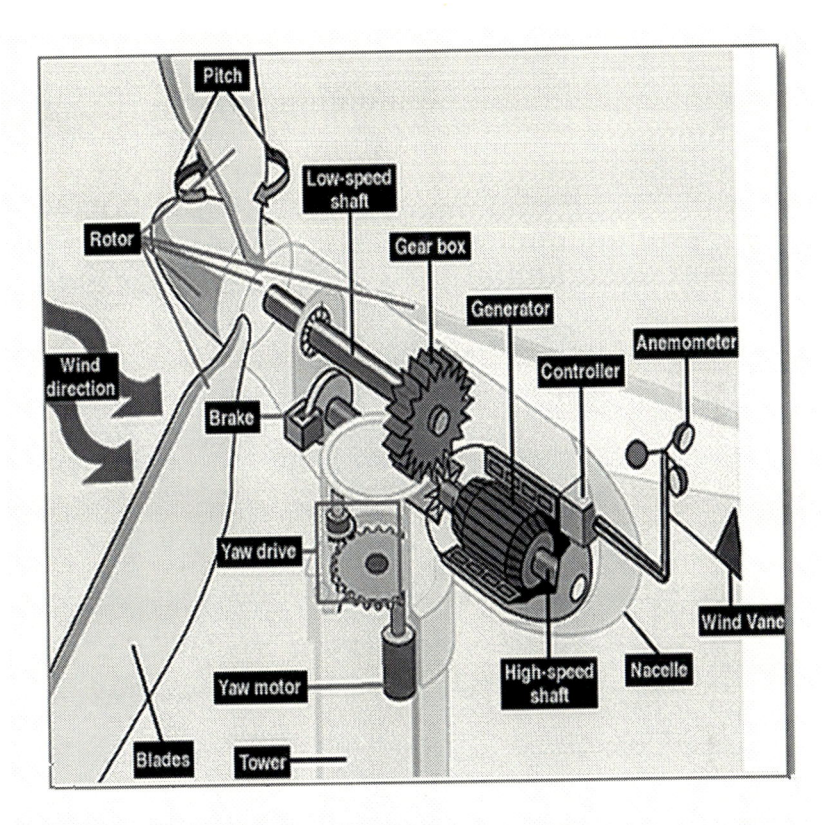

Figure 16. Wind Turbine Components[3]
[From the Energy Efficiency and Renewable Energy (EERE) program of the U.S. Department of Energy]

Wind turbines are constructed to tie into the home or business electrical system. The amount of electricity generated naturally depends on wind speed. In grid-tied systems, when wind speed is below the minimum required to operate the turbine – a point called the cut-in speed – the power grid functions as the supplier of electricity. As wind speed picks up, the electrical output of the wind turbine enters the home and the amount of power taken from the power grid decreases. When the turbine produces more power than the house needs, the grid not only ceases to provide electricity, it receives electricity from the home. When wind speed becomes too strong – a point called the cut-off speed – the wind turbine stops functioning. In many states, electricity from wind turbines that enters the power grid system is credited to the homeowner. So homeowners get paid for the excess electricity they produce. A power meter is used to monitor this interaction automatically.

Although wind turbines can be used by individual homes, they are scalable. A collection of large wind turbines can provide electricity for whole communities. In some states, wind farms with large wind turbines already produce enough electricity to service over 100,000 homes. And modern wind turbines are relatively quiet. At 900 feet away a 1.5 megawatt wind turbine that is 150 feet high with a base of 12-20 feet will produce less than 45 decibels (dB) [4] of noise. That is about the amount of noise produced by your fridge.

Wind turbines have one other formidable characteristic: their EPR is 50:1.[5] The EPR for wind turbines is surprisingly better than the EPR for energy systems using natural gas, coal, or even from oil pumped from the ground today.

Challenges for Wind Turbines

The main challenges for wind turbine technology are

>Limitations on Use
>Operating Range
>Noise
>Cost

Overcoming limitations on use is the main challenge. Many communities that allow wind turbines to be used require that they be situated on at least a ½ acre. Objections are raised to their use in cities because of noise, unsightliness, and because of the danger they pose to birds. The latter objection may be reduced as new models are introduced. Newer propeller type wind turbines move at a controlled, slow rate.

Wind turbines are also limited in operation, which not only affects where they can be used but how well they perform. Most wind turbines require rather high cut-in wind speeds before they begin operating - typically 7-10 mph. And many models have relatively low cut-off wind speeds of 30-40 mph. Almost as important a limitation is annual wind speed or how much wind a wind turbine can draw on.

Fortunately, newly designed wind turbines have an increased operating range. They can operate in smaller spaces at lower heights, in much lighter winds, in gusty winds, and in very strong winds. For example, the Finnish company Windside has a wind turbine that it claims will produce at least 50% more energy per year than its rival models thanks to a wider operating range. The company shows graphs of its top of the line wind turbines producing electricity at a wind speed of 2.5 m/s (5.6 mph) with no cut-off wind speed under European wind conditions. The reason for the increased operating range is a revolutionary style design that dispenses with propellers and uses a vertical axis of rotation.

So how much electricity can one of these Windside wind turbines produce? According to their data, their best turbine can produce 269 kWh/month when the wind speed is 2.7 m/s (6 mph) and 940 kWh/month when the wind speed is 6.3 m/s (14 mph).

What's more, because of their design, Windside's turbines are not only beautiful to look at, they appear to be harmless to birds. Despite their attractiveness, they do have an obvious drawback. Although other wind systems can be less expensive than solar panels, these systems are not. They are very expensive according to American Windside distributors I've talked to. Hopefully the cost will come down as they and their clones become more widely available.

Of the challenges for wind technology, improving operating range holds the greatest promise. One day it may be possible for wind turbines, or a related technology, to even harness the enormous energy in big storms and hurricanes. But even if they can increase their operating range and expand the energy resources they draw on, wind turbines will still require energy storage to compensate for times when the wind doesn't blow.

Solar Power

Not only was the 19th century the beginning of wind turbine technology for electrical power generation, but it was also the beginning of solar cell technology. Solar cells are based on another great scientific discovery, the *photovoltaic effect,* which was first described by French physicist Edmond Becquerel in 1839. The photovoltaic effect occurs when light strikes the junction of two dissimilar materials, such as a metal and a semiconductor; or a semiconductor with one type of impurity and a semiconductor with another type; or two different semiconductors. The light's energy is in the form of photons that dislodge electrons which are then free to conduct electric current.[6]

The invention that has allowed us to take advantage of the photovoltaic effect is the photovoltaic (PV) cell, which is also called a solar cell. PV cells capture the current of freed electrons for external use through metal contacts on the top and bottom of the cell. The power of a PV cell is approximately equal to the amount of current output multiplied by the voltage.[7] Connecting solar cells together in parallel or in series increases power. So solar cells are connected together in solar panels, which in turn are assembled into even larger collections called solar modules.

The manufacture of solar cells is now based on a mature technology with a very promising future, in part because solar cell efficiency has been steadily rising. Conversion rates for commercially available PV cells are now between 11 and 18%. The conversion rate is the percentage of the sun's radiant energy falling on a PV cell that the cell can convert into electricity.

As the manufacture of solar cells continues to evolve, they are not only getting more powerful, they are also getting more flexible. At least one company has already commercialized what is essentially a solar power roll – a set of solar modules attached to a material that is flexible enough to be rolled up like a thick carpet. This flexibility makes the solar power roll relatively easy to install and to use. The solar power roll has been successfully installed on a number of rooftops on commercial buildings in Los Angeles and overseas, particularly in Germany, which has a booming solar industry. Advances in solar technology promise to increase the flexibility of solar cells even more. One day soon we may see solar cells applied like paint.

CSP Technology

Solar technology is not confined to solar panels. There are other promising techniques for harnessing the power of the sun. An example is concentrating solar power (CSP) technology. There are four basic types of CSP systems currently utilized: Concentrating Photovoltaic Systems, Dish Systems, Parabolic Troughs, and Solar Power Towers which are also called Solar Towers.[8]

These four basic types of CSP all take advantage of a special property of parabolic mirrors: their ability to reflect the parallel rays from the sun in such a way that all the rays meet at a single point, called the focal point. This property is illustrated in Figure 17.

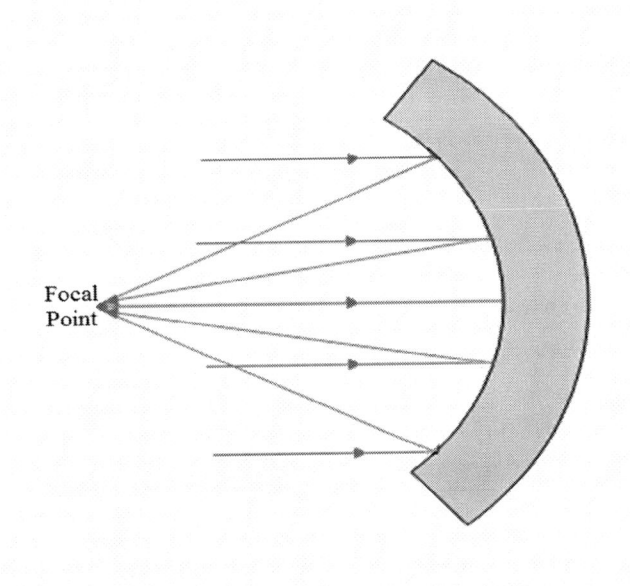

Figure 17. Parabolic Solar Concentrator[9]

The hub of a CSP system is the parabolic shaped structure called the **concentrator.** The concentrator is set up to track the sun continuously in order to ensure that sufficient sunlight is directed onto a **receiver.** The receiver is positioned at the focal point, where the reflected rays of sunlight are concentrated.

Although all of the four different CSP technologies that will be discussed here have solar concentrators and receivers, they have some clearly distinguishable differences that affect their cost, operation, and use.

Concentrating Photovoltaic Systems

A number of CSP systems use mirrors to gather light for PV cells. The main constraints on the CSP-PV systems for home use are expense and efficiency. Concentrating PV systems with mirrors require expensive, sophisticated tracking systems to keep them pointed at the sun. On cloudy days when sunlight is more diffuse, their ability to concentrate the sun's rays is significantly diminished, making them less effective than their non-concentrating competitors. To mitigate this problem, these concentrating PV systems have been located in sunny desert settings where they have been used for large scale production of electricity.

Solar Thermal Dishes

Not all concentrating solar power systems use the photovoltaic effect. Many, in fact, use the concentrated heat of the sun's rays to produce electricity. An example is a solar thermal dish system, like the one pictured in Figure 18.

In dish configurations, the **thermal receiver** is on the tip of an arm stretching out from the center of the parabolic panels as shown in Figure 18. In a solar thermal dish, the heat concentrated on the thermal receiver is absorbed by a heat transfer medium. Operating temperatures can reach 1,382°F (750°C).

Figure 18. Solar Dish[10]

The heat is conveyed to where it can be used by an electric generator. In solar dishes, the generator is driven by an engine.

Two types of solar thermal dish engines are used: the Stirling engine and the Brayton engine. In a Stirling engine, which somewhat resembles the internal combustion engine of a car, the heated gas is manipulated in four-cycles: compression, displacement, expansion, and displacement,[11] to move pistons[12] that turn a crankshaft which runs a generator.

The Brayton engine (Figure 19) like the Stirling engine converts heat to mechanical power by expanding and compressing a gas. The mechanical energy is then transformed into electrical energy. The

Brayton engine increases the temperature of gas already heated by the sun and compressed, using a **combustor,** which adds fuel and makes it a hybrid.

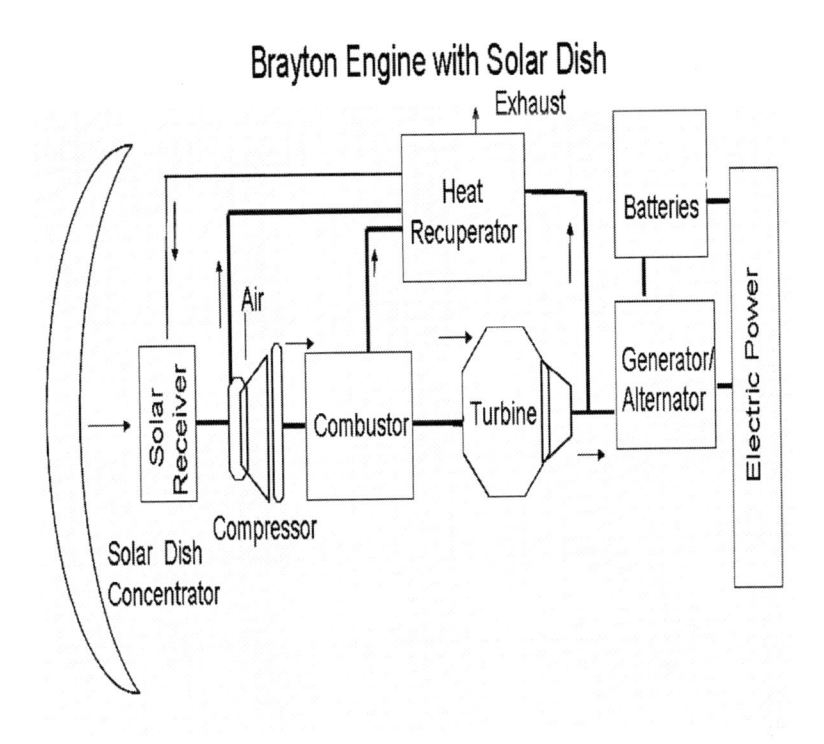

Figure 19. Schematic for a Solar Dish-Brayton System[13]

The gaseous mixture from the **combustor** is forced through a **turbine** where it expands. The expansion causes the turbine blade to spin, which turns the shaft of a **generator/alternator** which outputs electricity. A **heat recuperator** captures the waste heat produced by the turbine, combustor and compressor, and recycles the heat through the solar receiver and combustor, increasing the system's efficiency.

Solar Troughs

Another solar thermal CSP technology is the solar trough, which focuses the sun's rays along a solar **receiver pipe** rather than onto the smaller area of a thermal receiver.

Mechanical drivers slowly rotate multiple troughs so they follow the sun to maintain concentrated solar heat on the receiver pipe.

In solar troughs, synthetic oil acts as the heat transfer medium. The captured heat can cause oil temperatures to reach 734°F (390°C).

Figure 20.

Solar Troughs at Kramer Junction

in California [14]

The heat is transported in the oil through the trough pipes to a heat exchanger where water is boiled into steam which drives a turbine generator. Currently, gas-fired heaters are used as a backup so that enough electricity can be generated when the solar trough generators can't fulfill demand.

Solar Towers

A more exotic CSP technology is the solar tower which utilizes large, sun-tracking mirrors called heliostats to focus sunlight on a thermal receiver that sits atop a solar tower. The heliostat mirrors surround the tower at its base. [15] The original Barstow prototype called Solar I produced over 38 million kilowatt-hours of electricity during its operation from 1982 to 1988. Oil was used as the heat transfer medium. The steam that was produced ran a conventional turbine generator.

The second generation Barstow solar tower replaced the original oil transfer fluid with molten nitrate salt, which has superior heat transfer and energy storage capabilities. The use of molten nitrate salt has helped make the solar tower unique among solar electric technologies in its ability to efficiently store solar energy and dispatch electricity to the

BROADENING THE STATIONARY POWER BASE

grid when needed — even at night or during cloudy weather — without the use of storage batteries. In other words, instead of storing electricity in a battery, heat is stored in molten nitrate salt and can be converted to electricity when needed. It is one of the most economical means of storing energy yet devised.

The second-generation solar tower demonstrated a capability to deliver power to the grid night and day, except when there were several consecutive days of cloudy weather and its operation was reduced considerably. Figure 21 is a photo of a solar tower situated near Barstow, California.

Figure 21. Solar Tower[16]

The U.S. Department of Energy believes that "a single 100 megawatt solar tower with 12 hours of storage needs about 1000 acres of otherwise non-productive land to supply enough electricity for 50,000 American homes."[17]

In 2002 the Australian government gave the go-ahead to build a $550-700 million solar tower for commercial use. It is expected to provide 200 MW of electricity, enough to meet the needs of over 100,000 Australian homes. In 2005 Enviromission, the company that will build the solar tower, finally bought 10,000 hectare (around 25,000 acres) near Mildura for it. With a planned height of 3,300 feet, the tower would be the tallest manmade structure on Earth. The Australian solar tower is the most ambitious CSP undertaking in the world to date.

Working CSP systems have a field-tested peak conversion efficiency of 20-29%, which at first glance appears to be more efficient in producing electricity than home PV solar panels, but the annual efficiency of CSP systems even under ideal desert conditions is below that of the best conventional PV systems on the market.

Nonetheless, CSP systems, thanks to their size, produce much more power than home PV solar panels and are potentially capable of competing with coal and oil burning electrical plants in cost. Their main limitation is location.

Although CSP systems have been used for large scale electric power generation, I've wondered whether smaller versions might also be used for other applications. The CSP mirrors, thermal receivers, and pipes might be used to produce hot water in homes as well as heat and may even help to cool homes, if means can be found to store the heat for use at night and on cloudy days

Challenges for Solar Technology

The basic challenges for solar technology have been

Efficiency
Cost
Scalability
Limitations on Use

Efficiency has been for decades a major stumbling block. The solar cell technologies commercially available today, even the CSP PV systems, are limited in efficiency by their semiconductors. The semi-conducting material of solar cells only absorbs a small portion of the sun's spectrum of radiated energy. But what if it were possible to increase the absorption of sunlight, which is to say, what if we could increase the conversion efficiency of solar cells by using layers of different semi-conducting material, each of which can absorb a different portion of the sun's energy spectrum? The Jet Propulsion Laboratory (JPL) in California and its partners decided to find out and by 2003 they were getting impressive results. Their multi-layered semiconductor cells had a conversion efficiency of about 27%. But more was to come. In December 2006, one of those partners, Spectrolab, a Boeing subsidiary, announced that it had reached a conversion efficiency of 40.7% using a multi-layered concentrated solar cell similar to that used at JPL. In 2007, Spectrolab published its results in the journal *Applied Physics Letters*.[18] Shortly afterwards, the Department of Energy's National Renewable Energy Laboratory (NREL) published similar results in the same journal.

This achievement, needless to say, marks a revolutionary leap in conversion efficiency. What's more, the new record for conversion efficiency may soon be broken. Raef Sherif, director of concentrator products at Spectrolab, believes they will be able to produce photovoltaics with a conversion efficiency of 45-50% using concentrators.

So, have our problems with energy suddenly been solved? Unfortunately no. Present multi-layered, multi-junction solar cells are too expensive for practical individual commercial use. Though they may be more feasible for state governments than for individuals. But there is another problem for anyone wanting to scale up their use. Their materials are relatively rare and therefore limited in supply. In any case, the insights into conversion efficiency that this breakthrough has made possible, should bring us closer to a commercial solution. Variations on these multi-layered cells as well as other innovations in solar technology may soon realize the dream of low cost, relatively high efficiency solar cells. The potential candidates that currently show promise of improving the landscape of solar technology are thin film, quantum dot, organic, nanorod, and nanophotonic solar cells.

Storage Devices

Even if solar cells become more efficient and more economical, which would popularize the technology, their use is still limited to daylight hours of course. Furthermore, the peak power generated from sunlight occurs more than an hour before peak usage. A partial means of overcoming the limitation to daylight hours is to create a complementary hybrid solar-wind system. Such a system would extend the period in which electricity can be generated and would increase the amount of power available.

For solar cells and wind turbines to reach their full potential, however, we still need breakthroughs in energy storage just as we need such breakthroughs for the electric car. Today, energy is stored mainly in batteries. But conventional batteries used with solar cells and wind turbines exact a significant penalty. They not only make an alternative energy system significantly more costly (almost twice the cost), the use of batteries can reduce overall energy output as much as 30%. So the hunt is on for safe, economical storage systems with greater energy density and specific energy, and increased longevity and durability. Since energy storage is one of the key technologies affecting both the transportation and stationary power sectors, success in solving the problems with energy storage will have far reaching consequences.

Energy from Oceans and Rivers

Solar and wind energy are not the only potential sources of energy that are natural, clean, renewable, and environmentally friendly. Geothermal energy can be used to drive steam turbines. Similarly, the energy in water is also an option.

Thermocline Energy Systems

But how helpful would thermal energy from the ocean or lakes really be in reducing our oil imports and in reducing pollution? The Department of Energy estimates that each day the oceans absorb heat from the sun that equals "the thermal energy contained in 250 billion barrels of oil."[19] This is thousands of times more than the world currently uses. So the question is whether we could ever tap even a small portion of this energy.[20]

Unfortunately from an alternative energy perspective, the thermal energy that the ocean or lakes absorb has very low density, which makes exploiting any of it very difficult. But it is not impossible as George Claude, a French engineer, realized almost eighty years ago. Claude constructed an ocean thermal energy conversion (OTEC) system off the coast of Cuba back in 1929. He took warm tropical surface water from the ocean, put it into an evaporator, and changed the pressure, causing the water to vaporize. The resulting steam was run through a turbine to produce 22 kilowatts of electricity. Cold water was then piped up from lower ocean depths to cool the vaporized water so that the cycle could begin again. His project was abandoned because Caribbean storms kept breaking the pipe that collected cold water at lower depths. However, efforts to exploit thermal energy have continued, though at a slow pace.[21]

Currently, three types of OTEC systems are being tested: *open-cycle*, *closed-cycle*, and *hybrid systems*. The open-cycle system is essentially Claude's system, which turns lukewarm tropical seawater into steam. In contrast, the closed-cycle system uses ammonia or Freon to produce steam. Seawater in closed-cycle systems is used to change the state of the contained substance. Either liquid ammonia or Freon is selected because their boiling points are far below that of water. Surface seawater cyclically heats up the liquid ammonia or Freon as shown in Figure 22, vaporizing and expanding it so that it can spin the turbine of an electric generator. Then seawater drawn from a greater depth cools and liquefies the vapor.[22]

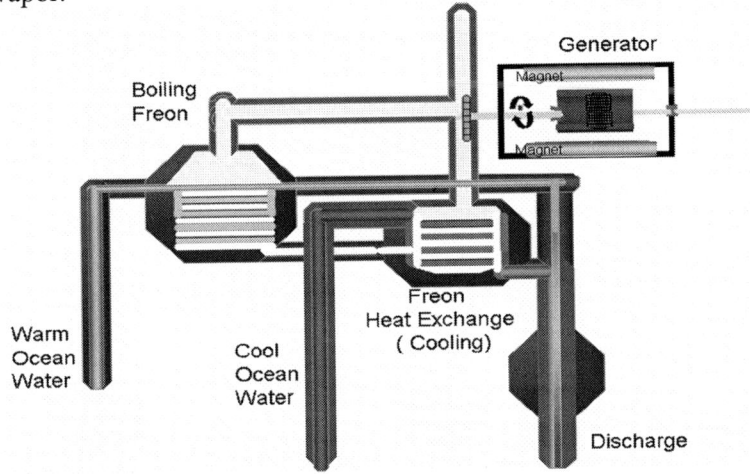

Figure 22. Closed-Cycle OTEC System [23]

The third type of OTEC system, the hybrid system, combines properties of both closed-cycle and open-cycle systems. All three OTEC systems are classified as thermocline energy systems to differentiate them from a simple ocean thermal conversion system which would use heated surface water.[24] Thermocline energy systems make use of the significant temperature differences that can exist between surface and deep ocean water.

Challenges for OTEC Systems

As thermocline systems, the three types of OTEC systems have some significant drawbacks which challenge those who want to make these systems commercially viable. The main challenges are

Cost
Limitations on Use
EPR / Efficiency

Costs are currently high because the technology is still at the demonstration stage. However, there are some limitations that can't be overcome no matter how many stages of development have been completed. Thermocline systems can't be located just anywhere. They need to be situated where the temperature at the water's surface is significantly different from the temperature deeper down. The Caribbean, where Claude constructed his thermal conversion system, is one of a few places that meet the criteria. The tropical waters of the Caribbean happen to meet and flow over deep currents from the arctic to produce a temperature difference of 35-40°F over a 1500-2000 ft vertical separation.

However, even with a 35°F temperature difference, the efficiency of conventional thermocline OTEC systems is quite low, about 2-3%. Part of the low efficiency is because 20-40% of the power generated has to be used for pumping water through pipes. Given just this energy expenditure, it isn't surprising to learn that a small-scale experimental Hawaiian OTEC system, the Kailua-Kona Test Plant, required roughly 150 kilowatts to produce only 50 kilowatts of electricity. The less-than-one EPR is clearly unacceptable. And that doesn't even include the energy required to create the OTEC equipment in the first place.

More recently, larger experimental OTEC systems off Hawaii and Japan have shown greater potential, suggesting that larger scale systems using new approaches might do more than hold their own. If they can operate at a net energy gain, they would have an important advantage over solar and wind. They could potentially produce electricity 24 hours per day, 365 days per year, with some seasonal variation. The bottom line is that a potential for thermocline energy systems exists, but breakthroughs will be needed.

Wave, Current, and Tidal Energy

Besides thermal energy, the ocean can also provide kinetic energy, along with rivers. The kinetic energy of rivers has, of course, been exploited through hydroelectric dams and waterfalls since Niagara Falls first provided electricity. I'm not advocating that we build more dams, but we could make more efficient use of what we already have. Although dams can produce the same amount of electricity twenty-four hours a day, our usage varies. During the night, electrical demand is generally lower. We could take better advantage of hydroelectric power than we currently do during this period of lower demand.

BROADENING THE STATIONARY POWER BASE   153

The potential for using the kinetic energy in waves, current, and tides has been much less exploited and is currently very promising. The Department of Energy estimates that the total power of ocean waves breaking on the world's coastlines could generate 2-3 million megawatts of electricity. In favorable locations, the wave energy density could average 65 megawatts per mile of coastline. Sixty-five megawatts is enough to provide electricity for around 32,000 homes.[25] Northern California and Oregon are particularly good candidates for generating wave, current, and tidal energy. The favorable conditions have led both California and Oregon to start wave power demonstration projects in conjunction with the Electric Power Research Institute (EPRI). The East Coast also has projects underway both in the ocean and in rivers.

The methods of capturing the oceans' mechanical energy are quite different from those used to capture thermal energy, and are quite diverse.[26] They range from *barges* (dams) that force the water through turbines, to buoys that translate the undulating motion of the waves into electricity, to oscillating water column systems that use the motion of waves to compress air within a container.

Challenges for Wave, Current, and Tidal Energy Technology

Their main challenges are

    Limitations on Use
    Reliability
    Cost

Although the technology for generating electricity from wave motion is still immature, it has already gone beyond demonstration projects in Europe, a promising sign. In November 2000, a Scotland-based company installed the world's first commercial system to generate electricity directly from surf. Known as Limpet 500 (Land Installed Marine Powered Energy Transformer), the Islay Wave Power Station can generate 500 kilowatts of power reliably, enough for about 400 Scottish homes despite the ebb and flow and changes in the level of the tides.[27]

The Scottish system is quite simple and elegant. It utilizes tidal waves to alter the water level inside a large, partially submerged concrete chamber built into the shoreline. Rising water forces air trapped in the chamber through a hole or holes and through a turbine. When the waves recede, the falling water level in the chamber sucks air through the turbine in the opposite direction. The key to the system's success is its use of the Wells turbine, named after its inventor, Professor Alan Wells of Queen's University. The Wells turbine cleverly rotates its blades in the same direction regardless of the airflow direction. This characteristic makes the Wells turbine perfectly suited for an ebb and flow system. The turbine spun by the air drives a generator that produces electricity. The UK remains a major center for wave, current, and tidal energy projects.

Along with these older ideas, newer ideas and devices for harnessing waves are cropping up. One intriguing example, invented by a California teenager Aaron Goldin in 2004, used a gyroscope paired with an electrical generator to extract power from ocean waves.[28]

Can Clean Alternatives Provide What We Need?

A crucial question in considering solar, wind, and ocean energy is whether they can produce enough energy to replace hydrocarbons in the transportation and stationary power sectors. According to Caltech professor Nate Lewis,

> "For a 10 percent efficient photovoltaic system, and the latest systems are 15 percent or better, we could supply all the United States' energy needs with a square of land some 400 kilometers on a side....this would cover the Texas and Okalahoma panhandles, part of Kansas, and a wee slice of Colorado" [29]

Others have calculated that the Earth receives around 10,000 times more energy from the sun than human civilizations consume today. [30] Wave and wind energy are less abundant than solar energy but according to Lewis we have enough wind energy to provide 100% of our needs at current rates of consumption. As for untapped wave, current, and tidal energy, according to the Electric Power Research Institute (EPRI), these sources of energy have the potential of providing at least 7% of our total present U.S. needs.[31] These calculations take into account currently available energy conversion technology.

Viewed from a wider perspective, in pursuing solar, wind, geothermal, and wave technology, we are setting off in a new direction in which we use science to bring ourselves into greater harmony with Nature, rather than using science to produce unnatural substances from hydrocarbons. By choosing to pursue cleaner, renewable energy, we can improve our health, environment, economy, and image.

15

A Parallel Path to the Future

At present there are few energy sources versatile enough to replace oil. Ethanol and methanol are two candidates, if sufficient quantities could be produced. Another is hydrogen, which like methanol may be used in applications from electric cars to home electric power generation. Hydrogen has been the top option of the Bush administration since 2001 and the top option of alternative energy advocates who believe our current Age of Transition will end in the Age of Hydrogen. So what are the prospects of transforming our current infrastructure into one that uses hydrogen?

A Hydrogen Energy Infrastructure: Challenges

Each of the major alternatives to oil has challenges to its widespread use. The biggest challenge for hydrogen is evolving our current energy infrastructure into one that uses hydrogen. As currently proposed by the government, a hydrogen infrastructure will require several interconnected system components as pictured in this U.S. Department of Energy diagram.

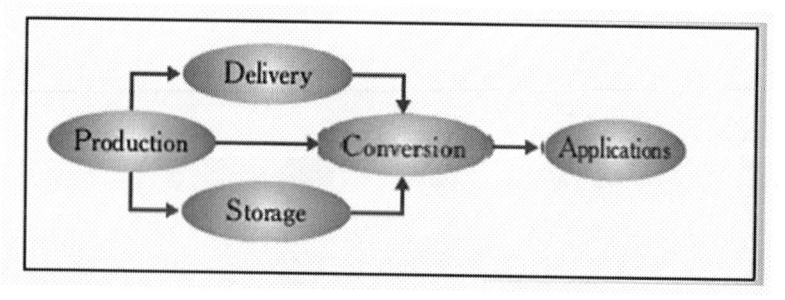

Figure 23. Hydrogen Energy Infrastructure[1]

A PARALLEL PATH TO THE FUTURE

What is involved in creating a whole new infrastructure? And where are we in that process? A look at the basic components that could make up a hydrogen infrastructure today can provide some answers.

16

The Hydrogen Internal Combustion Engine

Of the components that could be used to form a new hydrogen energy infrastructure, the most impressive, visible, and mature are the conversion devices, which convert the energy in hydrogen to perform useful work. In its *National Hydrogen Energy Roadmap*, the U.S. Department of Energy has identified two basic categories: combustion and fuel cell. In the transportation sector the main combustion conversion device is the hydrogen internal combustion engine (ICE) while the main fuel cell device is the polymer electrolyte membrane (PEM) fuel cell. Of the two, the hydrogen ICE is closest to the engines that we already see on the road.

The hydrogen ICE is similar to the gasoline ICE. So understanding how the gasoline car operates (Chapter 1) gives you a good idea of how a hydrogen ICE car operates. Surprisingly, getting an ICE engine to run on hydrogen isn't really that difficult. The Reverend William Cecil created a prototype around 1820. In a paper presented to the Cambridge Philosophical Society, Reverend Cecil described how he had burned a mixture of hydrogen and air, and on letting it cool and heat up again, induced changes in pressure in a cylinder that drove a piston up and down.

Some sixty years later, Nicolas August Otto talked of using "gaseous fuels" to run his four-stroke engine. Given the combustible gases available back then, it is possible the gaseous mixture he first experimented with contained hydrogen gas.

Otto expressed a preference for mixtures other than gasoline and air but changed his mind after Gottlieb Daimler's improvement of the carburetor, which reduced the danger of uncontrolled gasoline explosions. But the improved carburetor wasn't the only reason that gasoline prevailed. A review of hydrogen's characteristics and properties can provide necessary background for understanding not only why gasoline prevailed, but what challenges hydrogen use poses in modern cars, and what solutions have been found so far.

Elementary Properties

Hydrogen is the simplest, lightest, and most abundant element in the universe. It makes up about 75% of the mass of all visible matter. But the Earth is not the universe. On our planet hydrogen ranks only ninth in order of abundance among elements. It makes up less than 1% of the weight of the Earth's solid crust, but is abundant in the world's rivers, lakes, and oceans, and in biological organisms.

Water exists on Earth because of hydrogen's strong tendency to combine with other receptive elements, especially oxygen. Because it so easily combines with other elements, free hydrogen molecules are not abundantly found on Earth. Among natural forces, lightning is most likely to produce pure, uncombined hydrogen molecules by breaking the watery bond between oxygen and hydrogen. Mimicking nature, we can also produce hydrogen from water using electrolysis. But whereas we can capture and contain hydrogen, in nature pure hydrogen isn't likely to stick around. If not recombined with other heavier elements, hydrogen is so light, it will rise up and escape Earth's atmosphere.

Hydrogen molecules are light compared to other elements because the hydrogen atom contains so little matter. Normal hydrogen has only a single electron and a single proton. It has no neutrons. There are two isotopes of hydrogen that contain 2 and 3 times more matter respectively.[1] The isotope deuterium adds a neutron to the proton of hydrogen's nucleus, and tritium adds two neutrons. Both deuterium and tritium are rare.

Hydrogen is also unusual because of its boiling point. Liquid hydrogen doesn't exist naturally on Earth because it only becomes a liquid at -423°F (-253°C). The temperature at which an element is cooled enough to change from a gas to a liquid, or heated to change from liquid to gas, is its boiling point. It is a temperature of transition from one state to another. Hydrogen's boiling point is relatively close to absolute zero, the point at which all molecular motion ceases and far below the coldest natural temperature ever recorded on Earth.[2]

Even with human assistance, hydrogen is not easy to liquefy. At room temperature over 75% of normal hydrogen is orthohydrogenic, which means that each of the two protons in a hydrogen molecule has the same "spin." This property makes orthohydrogen unstable at temperatures near its boiling point.

To compensate, during the transition from gas to liquid, orthohydrogen changes into the more stable parahydrogen, which consists of a hydrogen molecule whose pair of protons have opposite spins. In the process of this transformation, the hydrogen molecule releases heat, which impedes liquefaction.

Challenges to Creating a Hydrogen ICE

So as the lightest element, hydrogen has low density, high buoyancy, high diffusivity, low boiling point, high flammability, and no toxicity. These properties translate into specific fuel characteristics within a hydrogen internal combustion engine (ICE):

Low Density
High Diffusivity and Radiating Speed
Small Quenching Distance
Low Ignition Energy
High Auto-Ignition Temperature
Wide Range of Flammability
No Toxicity

If hydrogen, not gasoline, had been selected earlier, these characteristics would have presented earlier engineers with a mixture of amazing benefits and probably insurmountable challenges. To see why, you need to understand these characteristics from an engineer's point of view.

Hydrogen Versus Gasoline

Density – The density of an element, whether in a gaseous, liquid, or solid state, is related to how close together its atoms are. In its gaseous state, the molecules of hydrogen are much farther apart than the molecules that compose liquid gasoline. Consequently, it takes more tank space to store hydrogen gas than gasoline – several orders of magnitude more.

Diffusivity and Flame Speed – Diffusivity relates to how easily and widely a substance can spread within a given time. Because hydrogen

THE HYDROGEN INTERNAL COMBUSTION ENGINE

molecules are exceptionally light, they diffuse easily, which makes them harder to contain than gasoline and makes their flame speed much faster than that of gasoline.

Quenching Distance – This is related to how close the flame of the combusted fuel in an ICE cylinder gets to the cooler cylinder wall before it is extinguished and is of the order of a millimeter. Hydrogen flames travel closer to the cylinder wall than gasoline flames before they extinguish, thanks to hydrogen's flame speed and greater capacity to diffuse. Unfortunately, the small quenching distance of hydrogen encourages detrimental hot spots on the cylinder.

Ignition Energy – This is the energy it takes to ignite a flame. Because it is highly combustible, pure hydrogen is easily ignitable by another agent. It catches fire more easily than gasoline.

Auto-Ignition – This is the temperature a substance self-ignites. Contrary to what might be expected, hydrogen has a relatively high auto-ignition temperature. Hydrogen's auto-ignition temperature is significantly higher than that of gasoline, which means gasoline self-ignites much more easily than hydrogen.

Range of Flammability – Hydrogen can be mixed with air in various concentrations from relatively highly concentrated to quite low and still ignite. This characteristic gives hydrogen a wider range of flammability than gasoline.

Toxicity – Hydrogen, unlike hydrocarbons, is non-toxic whereas gasoline fumes are highly toxic.

Premature Ignition Problems, Leakage, and Pyrolysis

How do hydrogen's characteristics as a fuel pose a challenge to automotive engineers? Hydrogen internal combustion engines are prone to premature ignition. This problem was widely reported in the late 1970s and the 1980s when hydrogen ICE buses were adopted by some cities, particularly in California. Premature ignition occurs when the fuel mixture in the combustion chamber ignites without the spark plug spark. The result is an inefficient, rough running engine.[3] Backfire conditions develop if premature ignition occurs near the fuel intake valve and the resultant flame travels back into the intake or induction system.

The hot spots that form on the cylinder because of hydrogen's smaller quenching distance serve as the main sources of unintended ignition. Hydrogen's lower ignition energy, diffusivity, and wider flammability range contribute further to the potential problem. And hydrogen's diffusivity makes leakage into the intake or induction system and the crankcase more likely.

In addition, hydrogen engines are more subject to pyrolysis. Pyrolysis of oil, chemical decomposition brought about by heat, is an ancillary result of pre-ignition. Pyrolyzed oil can enter the combustion chamber from the crankcase, through the crankcase ventilation system or through the intake manifold. The intake manifold is the part of the automotive system through which the fuel/air mixture is supplied to the cylinders.

Storage and Power Issues

The low density of hydrogen gas makes it difficult to store enough on board a car to drive very far. Having low energy density can also reduce power output. Changing hydrogen gas to liquid to improve storage capacity and energy density is also problematic. It is expensive not only monetarily, but also in terms of energy expended because of the need to cryogenically cool the hydrogen to -423°F (-253°C).

The Issue of Flammability

Another disadvantage of hydrogen is the potential danger it poses in a confined space. Hydrogen can cause asphyxiation by displacing oxygen. The victim is likely to be unaware of what is happening because hydrogen is odorless, colorless, and tasteless.

The Benefits of the Hydrogen Internal Combustion Engine

With such disadvantages, you might well ask why anyone would consider using hydrogen to run a vehicle. The answer is its tantalizing benefits. Engine efficiency is measured by the completeness of combustion at operating temperatures. Hydrogen's high diffusivity facilitates formation of a uniform mixture of fuel and air. Because hydrogen mixes more uniformly with air than gasoline, a hydrogen-air mixture leaves far less unburned fuel than gasoline does.

To get an idea of how superior hydrogen combustion is to gasoline combustion, you need to compare hydrogen's stoichiometric ratio (the ratio of fuel to air at which the fuel is completely consumed) to the stoichiometric ratio of gasoline. The air-to-fuel ratio for gasoline is 15:1 or 15 grams of air per gram of gasoline. In comparison, the stoichiometric ratio for an air-hydrogen mixture is 34:1 or 34 grams of air per gram of hydrogen. But thanks to hydrogen's wide range of flammability, hydrogen engines can do even better. A hydrogen mixture can run lean. A lean mixture is one that can be completely used up using air-to-fuel ratios greater than the stoichiometric ratio. Hydrogen is almost completely combustible at air-to-fuel ratios anywhere from 34:1 (stoichiometric) to 180:1. Such a lean-operating fuel is rare. And that is only part of the good news.

The high auto-ignition temperature of hydrogen allows greater compression to be used in a hydrogen ICE than in a gasoline ICE. Higher temperatures increase compression. Higher compression contributes to higher thermal efficiency. Higher thermal efficiency translates not only into more complete combustion, but greater power. Because of the greater possible compression, a hydrogen engine can approach the thermodynamically ideal engine cycle.

Another distinct advantage of having a fuel with a wide range of flammability is that the fuel mixture can easily be varied to meet different driving conditions or loads. Varying mixture ratios for hydrogen cars is similar to the strategy used by conventional diesel engines and contrasts markedly with what is possible for conventional gasoline engines. Because of its much narrower range of flammability, gasoline's fuel-to-air ratio must be kept within a narrower range throughout the driving range despite the lower fuel inefficiency that may result. Its narrow range of flammability is a major reason why a gasoline ICE has more trouble reaching the point at which its fuel is completely consumed than a hydrogen ICE does.

Overall, hydrogen is also less dangerous than gasoline. Hydrogen is far less dangerous if combusted in an unconfined space than is gasoline. Vapors from a gasoline fire, besides their high toxicity, can easily spread laterally, ignite from heat and involve the car interior. And the vapors can linger for quite a while due to gasoline's heavier molecules. Thanks to hydrogen's lightness and buoyancy, an unconfined hydrogen fire, besides being non-toxic, shoots straight up and quickly disperses.

Overcoming the Disadvantages

So the use of hydrogen in an internal combustion engine has both advantages and disadvantages that need to be understood in order to modify existing gasoline ICE designs. Since the main problems occur when hydrogen is ignited prematurely, this is what most engineers have focused on to design a successful hydrogen ICE. More specifically, they have looked at ways of getting fuel into an engine cylinder for combustion. The center of attention has been fuel injection.

In modern conventional fuel injection systems, whenever the gas pedal is depressed, the **throttle** opens up enough to allow more air into the engine's cylinders. Simultaneously, the **engine control unit** (ECU), part of the car's computer, aided by **air sensors**, opens the fuel injector to inject fuel into the engine cylinders. The amount of time the fuel injector stays open determines the amount of fuel supplied to the engine. This period of time is called its **pulse width**. The ECU controls the timing of openings and closings by issuing commands to energize or de-energize an **electromagnet**. The electromagnet when energized moves a plunger that opens the fuel injection valve, allowing the pressurized fuel to squirt out through a tiny nozzle. The nozzle is designed to atomize the fuel, that is, convert the liquid fuel into a fine mist so that it can burn easily.

As new engines were designed, a fuel injector was provided for each cylinder. This is called **port fuel injection**. With the advent of individual injectors for individual cylinders, a pipe called the **fuel rail** was added that supplies pressurized fuel to all of the injectors. The injectors were mounted in the **intake manifold**[4] so that they could easily spray fuel directly through the individual intake valves as they opened.

The power of fuel injection systems for conventional cars is that they provide more accurate fuel metering and quicker response than a carburetor, which is why they eventually replaced the carburetor. The value of the port fuel injection systems for hydrogen cars is their ability to better time the fuel injection with the air intake during the intake stroke. The port injection system can easily inject hydrogen into the intake manifold after the beginning of the intake stroke when the probability for premature ignition is reduced. Because air can enter the cylinder before the hydrogen, at the beginning of the intake stroke, it can naturally dilute hot residual gases and cool any existing hot spots.

A side advantage of the port fuel injection systems for hydrogen cars is that the interaction of the accelerator pedal, ECU, and sensors with fuel injectors can be used to vary the ratio of fuel to air in accordance with the power demands. To enhance this effect, the throttle remains wide open for hydrogen cars while varying amounts of hydrogen gas are injected. For conventional cars, in contrast, the throttle is not always wide open. Air flow is more restricted; the interactions of the accelerator pedal, throttle, ECU, sensors, and fuel injectors are designed to reduce variability in the ratio of fuel to air.

Because of their characteristics, port fuel injection systems are preferred for hydrogen ICE cars in order to mitigate irregular combustion due to pre-ignition and backfire. However, while quelling backfiring, port fuel injection does not completely solve the problem of premature ignition. Engineers have been forced to look at additional automotive system modifications to reduce that problem.

The modification that has proved most helpful is thermal dilution. Thermal dilution techniques, such as **exhaust gas recirculation** (EGR), curb pre-ignition by reducing the temperature inside the engine block. As the name implies, an EGR system re-circulates a portion of the exhaust gases back into the intake manifold. A recirculation of 25-30% of the exhaust gas gets rid of pre-ignition because it reduces the temperature that breeds hot spots.[5] As a bonus, it reduces the peak combustion temperature, which reduces NOx emissions. But, as often happens, there is a trade off. The power output of the engine is reduced when EGR is used. The reason is that the exhaust gas dilutes the fuel mixture that is drawn into the combustion chamber. You seldom get something for nothing.

A completely different approach to solving the hydrogen ICE's problems is to revamp the engine so that the combustion chamber is disk-shaped. While this is the most effective approach, it is also the most expensive.

The main effects of a disk-shaped combustion chamber are to reduce turbulence within the combustion chamber, increase the flame's distance from the cylinder walls, and widen the range of flame speeds. The first two effects reduce the problems they were meant to address, premature ignition and backfire. But widening the range of flame speeds introduces a new problem; it affects lean fuel ignition, reducing power when lean mixtures are used. To combat this new problem requires some modifications to the ignition system as well.

Innovations in Ignition Systems

To improve the ignition system dual spark plugs are used. They help lean mixtures ignite at an increased distance from the cylinder wall. The lean mixtures affected have stoichiometric ratios that range between 130:1 and 180:1.

Besides dual spark plugs, all hydrogen cars run better with spark plugs that are cold-rated and with spark plug tips that don't have platinum. Unlike gasoline engines, which can use hot rated spark plugs to reduce carbon deposits, engines that use hydrogen have no need for them.

Moreover, cold-rated spark plugs cool the spark plug tip more quickly than hot rated spark plugs, reducing the chance of pre-ignition problems.[6] So, cold-rated plugs are naturally preferred for hydrogen engines. As for platinum-tip spark plugs, they are supposed to improve the mpg of conventional cars. But they harm hydrogen cars because platinum is a catalyst[7] whose interaction with hydrogen increases the chance of premature ignition.

Reducing Premature Ignition through Crankcase Ventilation

Additional measures that have been found to decrease the probability of pre-ignition are (1) the use of two small exhaust valves as opposed to a single large one, (2) a revamping of the cooling system to provide more uniform flow of coolant to all locations that require cooling.

Tackling Other Problems

While pre-ignition and backfiring are the major problems, there are others that have to be tackled as well, like combustion chamber leakage. Modifications need to be made to prevent hydrogen from entering the crankcase compartment below the combustion chamber where the crankshaft resides.

If unburned hydrogen is allowed to accumulate within the crankcase compartment it can ignite, causing a sudden rise in pressure that can damage the crankshaft. Ventilation is one of the simpler methods used to

THE HYDROGEN INTERNAL COMBUSTION ENGINE

significantly reduce the problem. If unburned hydrogen enters the crankcase, ventilation can dilute it, and if the hydrogen ignites, ventilation can immediately reduce pressure.

Another reason that ventilation is important is that it reduces the accumulation of water vapor. Water vapor accumulates as a product of the hydrogen/air combustion process. It can mix with oil in the crankcase, reducing the effectiveness of lubrication. Less lubrication results in more engine wear. Proper ventilation can reduce this problem by sucking out the water vapor.

Another problem that must be tackled is emissions. The only product resulting from the combustion of hydrogen with pure oxygen is water.[8] However, it is currently cheaper to use air, which contains oxygen and nitrogen, rather than pure oxygen in the combustion of hydrogen, and that poses an emissions problem. The combustion of hydrogen with air produces oxides of nitrogen[9] due to the high temperatures generated within the combustion chamber. The magnitude of the emissions problem depends upon

(1) air to fuel ratio
(2) ignition timing
(3) engine compression ratio, and
(4) speed with which the piston moves

In addition to oxides of nitrogen, traces of carbon monoxide and carbon dioxide can be present in the exhaust gas, due to lubrication oil that seeps into the combustion chamber and is burnt. If the engine is burning oil, which may occur in any ICE after years of engine use, pollution emissions can be even more significant. Retaining a catalytic converter can make a difference and provides extra insurance that the amount of pollution vented into the atmosphere is low.

The Struggle

Given the solutions that exist to counter backfire, leakage, emissions, and pre-ignition, designers should be free to capitalize on hydrogen's advantages, right? Yes, but not without a struggle. Hydrogen engines should achieve at least 15-30% better power output than gasoline engines, but solutions to the backfire and pre-ignition problems can prevent hydrogen ICEs from achieving their expected power potential. Moreover, since one of the reasons for using hydrogen is to lower exhaust emissions, hydrogen engines are not normally designed to run at a stoichiometric air-to-fuel ratio but instead use about twice as much air as theoretically required for complete combustion. At the higher air-to-fuel ratio of 68:1, the formation of NOx is reduced to near zero, but the power output is reduced to about half that of a similarly sized gasoline engine.

To overcome these constraints, hydrogen engine designers have increased engine size and equipped their hydrogen engines with turbochargers or superchargers. Such design decisions have brought manufacture prototype hydrogen engines close to their potential.

Ford's H2RV Hydrogen Internal Combustion Engine

One example of how far manufacturers have come is Ford's Hydrogen Hybrid Research Vehicle (H2RV), based on Ford's Model U concept car[10] and on the Ford Focus wagon.

Ford engineers have successfully created a hydrogen ICE that is easy to drive, goes over 200 miles before requiring refueling, and gets 45 miles per kilogram (kg).[11] A kilogram of hydrogen is equivalent to 0.28 liquid gallons. The H2RV has near-zero emissions. Carbon dioxide alone is reduced by 99% from that produced by the conventional Ford Focus wagon. And there have been no pre-ignition or backfire problems reported. Here are some of the specific reasons for Ford's success.

Fuel Storage

The H2RV is capable of carrying up to 7 kg of hydrogen on board. At 45 miles per kg,[12] this would allow it to go 315 miles per fill up. It currently carries around 5 kg, however, according to Ford. To carry sufficient fuel, the H2RV uses pressurized hydrogen gas, which requires a fuel tank that is made of a 3-millimeter aluminum pressure barrier with a carbon-fiber structural casing. The fuel tanks are rated to an operating pressure of more than 10,000 psi.

Fuel Injectors

For the H2RV, Ford uses dual fuel rail, hydrogen-tolerant fuel injectors similar to those that can be found in a car modified for hydrogen at the University of California, Riverside Center for Environmental Research

THE HYDROGEN INTERNAL COMBUSTION ENGINE 169

and Technology. The fuel injection system is designed to vary according to power needs, to run lean (use more air) as much as possible, and to precisely time the injection of fuel during the four-stroke cycle of the engine to avoid pre-ignition problems. The **intake manifold,** through which hydrogen gas reaches the engine, has a 1½" diameter. A ¼" tube transports pure hydrogen to within an inch of a cylinder's **intake valve**. The closeness of the tube end point to the intake valve minimizes the amount of hydrogen that would be in contact with air and consequently reduces damage to the intake valve, if pre-ignition occurs.

Much like a diesel engine, the Ford hydrogen ICE runs unthrottled while under way, with air to fuel mixtures as lean as 86:1 during highway cruising speeds. As a result, Ford's version of the hydrogen ICE can reach an overall efficiency of 38 percent, which is at least 26% better than a gasoline engine.[13] Ford's design makes it easy to meter the hydrogen flow and vary the amount of hydrogen dispensed to the engine according to the power demands. A side benefit of the design is that any occurrence of premature ignition in one cylinder is isolated. It won't cause premature ignition in other cylinders.

Supercharger

To increase the amount of hydrogen in the engine cylinders, Ford uses the Vortech supercharger. The supercharger provides nearly 13 psi of boost on demand, which increases the amount of air entering the engine by about 100%. Because of the increased air in the cylinder, more hydrogen can be injected into the cylinder without decreasing the air-to-fuel ratio below the desired level. Using a supercharger or turbocharger is not the only way to increase fuel density. It can also be increased if hydrogen is mixed with other fuels such as natural gas.

Hybrid Electric Transmission

Ford further increases the H2RV's fuel efficiency and reduces emissions by using hybrid technology. The H2RV power train features a **Modular Hybrid Transmission System (MHTS)**. The MHTS provides the H2RV with the standard regenerative braking function that reclaims energy otherwise lost to friction and stores it in a 300-volt, air-cooled battery pack. MHTS technology was designed by starting with the base conventional transmission. The design makes minimal modifications to reduce costs compared to other hybrid systems.

Cooling Systems and High-Compression Pistons

Like its predecessor the Model U, the H2RV uses intercooling, a dual-stage process, to reduce potential hot spots in the engine cylinders. After leaving the supercharger, the intake air passes through a conventional intercooler.

To improve engine performance, the engine is optimized to burn hydrogen with 12.2:1 high-compression pistons. The flattened, high compression pistons work with the superchargers to compress air flowing into the engine, consequently increasing the power from each explosion in each cylinder.

Emissions

Ford's design of the H2RV put it in the category of near-zero emission vehicles. Ford claims that new research into catalysts may soon reduce emissions below the ambient pollutant levels of many cities. If that happens, the exhaust of the H2RV would be cleaner than the air it takes in and would open up an intriguing possibility. The H2RV might actually become an air cleaner. Instead of just reducing its own pollution levels, it could reduce pollution already in the ambient air.

Engine Management Software

Through its engine management software, the Ford H2RV is able to vary hydrogen fuel injection according to power needs. The engine control computer (ECC) determines the amount of desired fuel, principally from a position sensor connected to the accelerator pedal. The accelerator pedal sensor sends a fuel demand signal to the ECC. The ECC controls how long it will hold an injector open based on this signal and the input from several other sensors including an oxygen sensor, voltage sensor, engine speed sensor, and manifold absolute pressure sensor. The latter sensor monitors the pressure of the air in the intake manifold.

The Other Two Legs of the Stool

The H2RV has the potential to both improve the environment and drastically reduce our dependency on oil, and it could do so now. What keeps the H2RV prototype from being commercialized is described by

Dr. Gerhard Schmidt, vice president, Ford Research and Advanced Engineering,

> "What we are lacking are the other two legs of this three-legged stool – a fueling infrastructure for hydrogen, and uniform laws and regulations that will allow its use across the nation."[14]

17

The PEM

If we do build a hydrogen based infrastructure in the transportation sector, we are likely to enter a period in which two types of hydrogen car will compete, the hydrogen internal combustion engine (ICE) car, which has just been described, and the hydrogen fuel cell car. The basic hydrogen ICE car, as you are now aware, is an adaptation of the gasoline ICE car. The fuel cell car, on the other hand, is a variant of the electric car. Both not only have the potential to end our dependency on foreign oil and clean up our environment, but they also have the potential to generate new industries with hundreds of thousands of domestic jobs. Of the two types of hydrogen car, the fuel cell car is likely to take longer to become established because it is more experimental and revolutionary in design.

Yet even though the fuel cell car is more experimental, it is, like so many other inventions associated with the twentieth and early twenty-first centuries, a product of the nineteenth century. Credit for the invention of the fuel cell is generally given to Sir William Grove, a nineteenth century Welsh judge and gentleman scientist, and is based on his remarkable discovery of inverse electrolysis in the early 1840's.[1]

Inverse electrolysis is the reverse of electrolysis. Whereas electrolysis splits water into hydrogen and oxygen, using electricity, inverse electrolysis combines hydrogen and oxygen to form water, and in the process produces electricity.[2]

Grove's essential insight was that if an exact inverse process existed, it would have to act much like a conventional battery connected to an electric circuit.[3] Grove called his fuel cell a "gas battery."

His insight arose from knowledge of electrolysis, catalysts, and the known characteristics of batteries. Grove knew, for instance, that certain catalysts can strip hydrogen of electrons and that electrons can be induced to flow through a circuit with electrodes on either end. Figure

24 represents a simple experiment Grove concocted to explore the possibility of inverse electrolysis. [4]

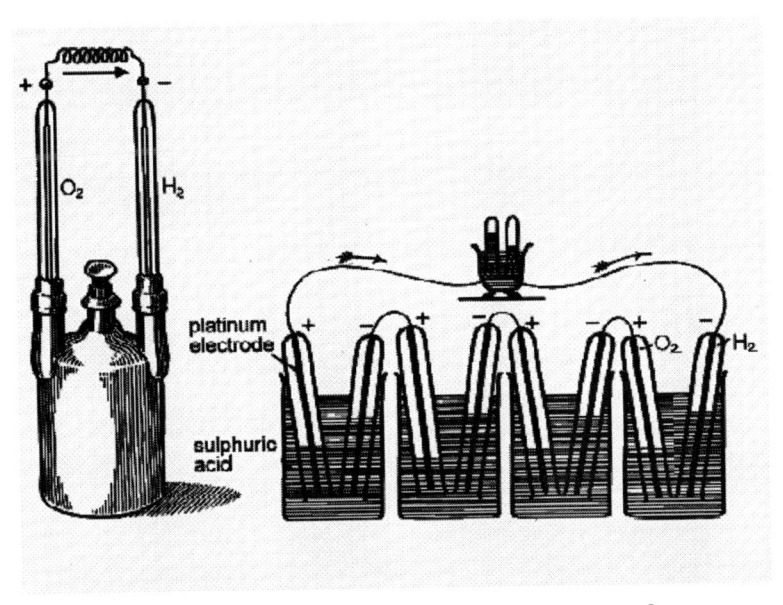

Figure 24. Grove's Fuel Cell Configuration[5]

In the drawing, he has a set of fuel cells connected to a separate apparatus used in electrolysis. The fuel cells produce the electricity that the separate apparatus uses for electrolysis.

Each fuel cell is composed of two tubes. One contains hydrogen gas (H_2), and the other contains oxygen gas (O_2). Both tubes in one of these fuel cells are immersed in an electrolyte comprised of sulfuric acid (H_2SO_4) and water. The sulfuric acid is shown partially filling the tubes from below. The hydrogen and oxygen gases each rise to the top of their tube above the sulfuric acid. Each tube also encloses a platinum electrode strip which attaches to the circuit that extends outside the tube. The platinum strip in the hydrogen tube is called the anode and the one in the oxygen tube is called the cathode.

When a hydrogen molecule (H_2) comes in contact with the platinum strip anode, it splits into two separate protons (H^+) and two free electrons (e^-). The protons pass through the electrolyte to the cathode and the electrons go through the external circuit to the cathode. After the protons and electrons get to the cathode, they combine with the oxygen to form water.

In order to demonstrate that the fuel cell produced electricity, Grove could have connected it to any electrical device. But in the 1840s there weren't many around! Grove did have an electrolysis apparatus, which he probably chose so he could demonstrate the fuel cell process and its inverse process at the same time.

The electricity that was produced by the fuel cell arrived at the separate electrolysis apparatus (seen in the background in Figure 24) where it split the water, producing hydrogen and oxygen. The oxygen is attracted to the positive (+) electrode. The hydrogen is attracted to the negative (-) electrode. Each gas bubbles up along its electrode and is confined in a tube. The volume of gas in the two tubes differs as shown in the illustration.

Subsequent experimenters have used a more efficient configuration of a fuel cell in which the electrolyte is sandwiched tightly between an anode and a cathode chamber. The effect is to allow ions to pass directly from the anode chamber into the cathode chamber through the electrolyte, while the electrolyte blocks electrons from using the same path. The proton exchange membrane (PEM) illustrated in Figure 25 is a prime example of this advanced thinking.

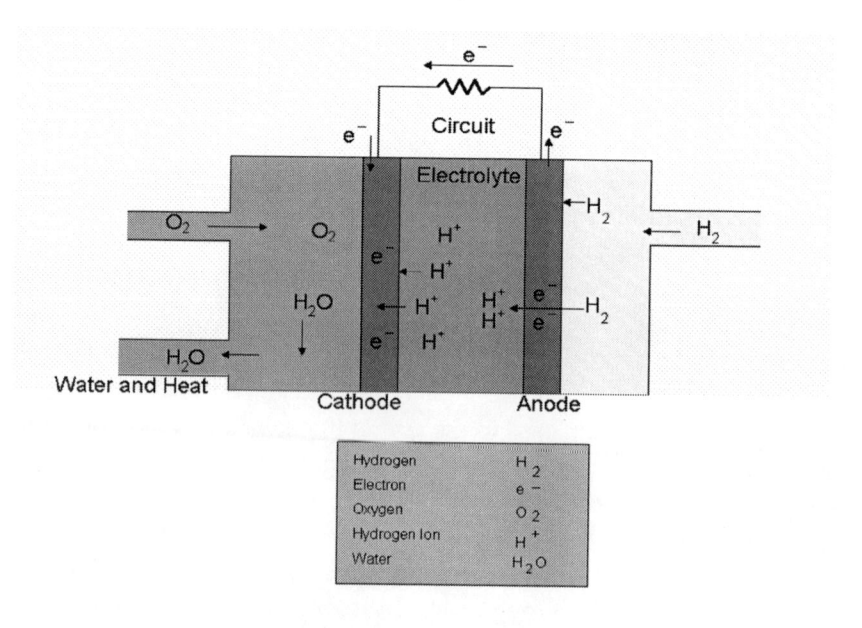

Figure 25. PEM Fuel Cell [6]

As with Grove's experimental configuration, electrons still flow from one electrode through an external circuit to the other electrode after performing work.

Although Grove saw his fuel cell as a "gas battery," a fuel cell does differ from a conventional battery in both obvious and subtle ways. In conventional batteries, the fuel is contained within the battery. In fuel cells, the fuel enters the battery from the outside. This difference gives the fuel cell a split personality. Provided with hydrogen, fuel cells produce electricity to drive a motor, which resembles the internal combustion engine's use of combusted gasoline to drive pistons. But the fuel cell resembles a battery in configuration, even though the fuel cell doesn't store energy.

The original single hydrogen fuel cell was not by itself capable of producing much electricity. A typical output might have been 6 amps and 0.7 volts. Not much could be done with a single fuel cell. But fuel cell researchers quickly learned how to increase the power by linking the fuel cells in series. Fuel cells arranged in series are called a fuel cell stack. The number of fuel cells in a stack varies, depending on the power needed. For instance, it would have taken 24 fuel cells with an output of 6 amps and 0.7 volts each to power a 100 watt light bulb.[7] With the invention of the fuel cell stack, fuel cells became a potentially important energy conversion device.

Proton Exchange Membrane (PEM) Fuel Cells for Cars

Since Grove's initial discovery, several types of fuel cells have been developed. All operate similarly to Grove's original fuel cells but not exactly. Differences are due largely to the electrolyte used, which is why fuel cells are classified according to their electrolyte.

Of the major types of fuel cells, all but two have electrolytes that require chemical reactions at such high temperatures that the fuel cells are not useable in cars. The two fuel cells that work best in cars to date are the **phosphoric acid fuel cell** (PAFC) and the **proton exchange membrane** (PEM) which is also called the polymer electrolyte membrane. Because of the time it takes PAFCs to warm up, their corrosive liquid electrolyte, and their relative bulkiness, PAFCs are a problematic choice for commercial grade cars so far, though they have been used in buses. The other fuel cell, the PEM, is much more suitable and is today the fuel cell of choice in the transportation sector. It is found in most prototype fuel cell cars. To fully assess its current status and future potential, you need to know a little about its development and how it works.

Early Challenges for the PEM

When in the mid 1950s American chemist Willard Thomas Grubb first came up with the idea of using a polymer electrolyte membrane as an electrolyte in a fuel cell, his company General Electric (GE) scoffed at the idea. Fuel cells were supposed to have electrolytes that are liquids, usually very hot liquids. Grubb's membrane was a solid, and frankly, the PEM fuel cell just looked weird. Its electrolyte looked something like ordinary kitchen plastic wrap.[8]

Although GE initially ignored Grubb's invention, its executives soon had reason to change their minds. No, GE did not suddenly decide to use PEM fuel cells to run cars. Oil appeared far too plentiful and cheap to consider an alternative. What changed GE's mind was the space race, which began on October 4, 1957 with the Soviet Union's successful launch of Sputnik. GE's executives recognized the potential of the PEM fuel cell to become an onboard power source for spacecraft. They resurrected Grubb's fuel cell and had their engineers set about making improvements. The first challenge was to make the platinum adhere better to the membrane.

Less than a year later, GE chemist, Leonard Niedrach came up with a way to deposit the platinum catalyst onto titanium gauze and bond the gauze to Grubb's polymer membrane. This improvement guaranteed that the platinum catalyst would stick and that the PEM would generate much more electricity as a result.

With this breakthrough, GE sought a market for their Grubb-Niedrach fuel cell in the National Aeronautics and Space Administration (NASA) space program. By the mid-1960s they were successful, but not with the customer they had originally targeted. Both the U.S. Navy's Bureau of Ships' Electronics Division and the U.S. Army Signal Corps wanted to experiment with fuel cells as a source of electricity for personnel in the field, but they worried about the expense. While the cell was compact and portable, its platinum catalysts were expensive. GE's successful demonstration of the fuel cell's utility overcame the reservations of the U.S. Army Signal Corps and won GE a contract. What was more important, it got the attention of GE's intended customer, NASA, which decided to use PEM fuel cells in the *Gemini* Earth-orbit space program.

The future of the PEM looked very promising, but GE soon ran into real-world problems, the worst of which was internal cell contamination and leakage of oxygen through the membrane. The leakage contaminated the anode and generally diminished the capability of the fuel cell to generate electricity. The problem is not unlike the problem that the Pennsylvania researchers encountered with the microbial fuel cell several decades later. Consequently, Gemini 1 thru 4 flew with conventional batteries instead. GE redesigned their PEM cell, and the new model P3, despite malfunctions and poor performance, flew on Gemini 5. NASA continued to use GE's fuel cells for most of the remaining Gemini flights, despite GE's failure to solve its oxygen problem. But GE's difficulties slowed advances in PEM technology for years and motivated the development of a rival.

In the early 1960s, Pratt & Whitney (P&W) the engine division of what is now United Technologies Corporation (UTC) acquired another type of fuel cell called the alkaline fuel cell (AFC). Pratt & Whitney quickly concluded that the AFC could outdo the PEM. The AFC had a longer life than GE's PEM cells and could be produced more cheaply. It was also more reliable. Pratt & Whitney set about making it even more attractive by reducing its weight. Their innovations won the competition for future NASA contracts, and their AFC went on to fame in the Apollo program. The defeat for GE quickly cooled interest in PEMs, but left a lingering desire among some to improve it.

There was no question that the PEM fuel cell was extremely attractive – in theory. Because of its electrolyte, it has the lowest operating temperature of any fuel cell, about 176°F (80°C). The relatively low temperature broadens the scope of use and reduces expenses. Manufacturers don't require expensive containment and insulation structures. Though a solid, the flexible PEM electrolyte doesn't crack. It avoids the corrosiveness associated with liquid electrolytes and, being a solid, is easier to handle. Moreover, the oxygen leakage that was inhibiting the PEM's use was theoretically preventable. But for years most research focused on the cheaper AFCs and on the PAFC, which could run on hydrogen gas that contained a significant amount of impurities. Young researchers interested in fuel cells were warned away from PEMs and directed to conduct their research in a more promising area.

Breakthroughs in PEM Technology

The few companies and researchers who did persist in PEM research were considered crazy by those "in-the-know." PEM researchers were "tilting at windmills," wasting their money pursuing an impossible dream. GE was one of those companies on that "impossible mission" to find a cure for what ailed the PEM. But no one made any progress.

Then in 1983, over thirty years after it was invented, a little known company in Vancouver made a breakthrough. Ballard Research Company was able to create a PEM cell that could run on impure hydrogen gas without poisoning the catalyst. They seemed to have defeated the oxygen leakage problem.

Emboldened, Ballard sought to tackle other problems, in particular the cost of the platinum electrodes. In the late 1980s, Los Alamos National Lab and Texas A&M University joined them in their pursuit, and Los Alamos quickly began to get results. Following Los Alamos's lead, Ballard manufactured an ink composed of tiny (2nm diameter) particles of platinum attached to carbon particles. Ballard's researchers applied the ink to the electrodes and tested the fuel cell. With only 1/10 of the original platinum in the composite mixture, the researchers were able to make the fuel cell work. The cost of the electrodes was suddenly and dramatically lowered.

Finally those who had decided to defy the general wisdom were unlocking the power of the PEM. The Ballard researchers were elated, but the elation didn't last long. They soon discovered that traces of carbon monoxide from their impure hydrogen could contaminate the platinum electrode. Somehow the technique that allowed them to reduce the amount of the platinum had enhanced the effects of contamination. They were back to the original problem of platinum electrode contamination in a slightly different guise.

What could they do? They added another metal, ruthenium, to the platinum electrode. Ruthenium not only stopped the carbon monoxide contamination but it significantly increased the lifetime of the PEM fuel cells.

Buoyed by success, Ballard's researchers expanded their efforts. They began putting their fuel cells together to form mega stacks that dramatically increased output current. By the end of the decade the company, which had changed its name to Ballard Power Systems, had more than doubled the power density of its stacks and dramatically increased their output. Improvements continued throughout the 1990s. The power density went up from 70 kilowatts per cubic meter (kW/m^3) in 1989, to 175 kW/m^3 in 1991, to 350 kW/m^3 by 1993, and to almost 1 megawatt per cubic meter (MW/m^3) in 1995. By 1997 it had reached 1.8 MW/m^3, which was more than enough to run an automobile or even a bus. In June 1993, with the power density of its fuel cells climbing, Ballard unveiled a trial PEM fuel cell bus, the first, and in the same year agreed to a joint venture with Daimler-Benz[9] to produce several prototype fuel cell hydrogen vehicles (FCV), which they called the Necar (New Electric Car).

At last, the big car manufacturers, who previously had not just been indifferent but sometimes hostile to the idea of fuel cell cars, began to show an interest in the technology. In 1997 Ford joined Daimler-Benz as an investor. Daimler-Benz acquired a 20% stake and Ford a 15% stake in Ballard.

Subsequently, two new companies, joint ventures of Ford, Daimler-Benz, and Ballard, emerged: Xcellsis which specializes in integrating fuel systems with cell stacks to make fuel-cell motors; and Ecostar which supplies complete electric power trains, including motors and control systems.

With the increases in the power density of PEM fuel cell stacks came significant decreases in their size. By the year 2000 a PEM fuel cell stack the size of a small piece of luggage, and an electric motor, could replace the internal combustion engine in a standard car. And PEM fuel cell stacks of an even smaller size could easily supply the electricity for most household and office appliances.

Advantages of using PEM Fuel Cells

The advantages of using PEM fuel cells to power automobiles are enormous. The PEM fuel cell has no moving parts so its design is much simpler than an internal combustion engine (ICE) and consequently a fuel cell car requires less maintenance. And because they require fewer energy conversions than conventional cars, they can be more energy efficient than cars with ICEs.

The reduction in energy conversions is illustrated in Figure 26. In the figure, the ICE's energy conversions from the chemical energy of fuel are in the dashed-line box. Note that instead of the two energy conversions to thermal energy and mechanical energy of the ICE car, the fuel cell car has only one energy conversion to electrical energy. Furthermore, since the fuel cell does not have a thermal energy step, there is far less wasted heat energy.

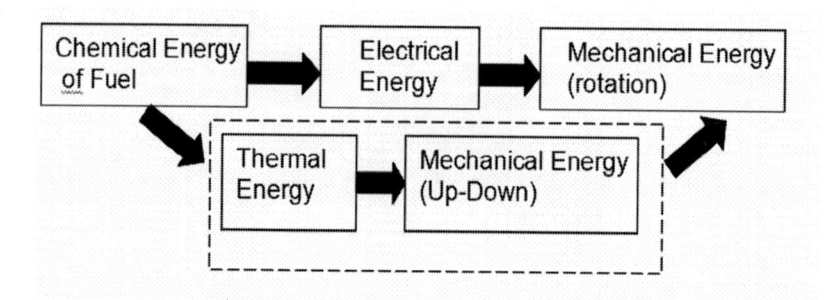

Figure 26. PEM verses ICE Energy Conversion Process[10]

Currently, the actual efficiency of PEM fuel cell cars is about 45% compared to the 20-30% efficiency of a conventional ICE. The main part of the exhaust from a hydrogen fuel cell car is water from the hydrogen fuel and the oxygen in air. Since the air that the fuel cell uses contains nitrogen too, there will be a small amount of pollutants from the nitrogen and will require a catalytic converter to clean it up. The fuel cell car's emissions are still far below that of a conventional car. Also, because the PEM fuel cell car doesn't require gasoline, it has no need of additives to boost octane or reduce knock. Nor does a fuel cell require lubricating oils. In short, it does not require so many of the poisons used to operate ICEs.

Continuing Challenges

The history of the PEM fuel cell is fairly typical of technological innovations. Many of the most revolutionary products begin as a basic research discovery or maverick idea with no obvious use. Often at first they are scoffed at because they are so different. But in time, potential uses are suggested and research intensifies. Obstacles can arise to discourage further development and impede progress, possibly for decades. Then, seemingly out of nowhere a breakthrough occurs and quickly substantial progress is made. The PEM is at the stage where it has the potential to revolutionize how we obtain our electricity in the transportation sector. Though it still has obstacles to overcome, the progress has been impressive, as you will see from the amazing Hy-Wire.

18
Hy-Wire Act

Some major car manufacturers including Daimler, Toyota, Ford, BMW, and Honda, make hydrogen PEM fuel-cell car prototypes. But the most revolutionary one to date is GM's **Hy-Wire**. From its first showing as a concept vehicle in 2002, the Hy-Wire has been a showstopper.

The strength of its design is underlined by the speed with which GM went from a rough concept to a full-blown prototype vehicle. The Hy-Wire's revolutionary design begins with the chassis on which the motor resides. This chassis, known as "the skateboard," is the core of the vehicle. Unlike the engines of conventional cars, the Hy-Wire's motor is tucked beneath the underbelly of the car within the skateboard. Its position and the general design have important implications. The skateboard, which is only 14.25 feet long, 5.47 feet wide, and 11 inches thick, can be fit with many different bodies. The same core vehicle can become a truck, an SUV, a compact sedan, a jeep, or any number of other recognizable models.

Mechanical locks secure the body to the skateboard, making car bodies easy to attach. Because the vehicle's engine is within the 11-inch thick detachable skateboard, there is no motor in front. In the absence of a front motor, the windshield reaches down to the level of the driver's waist, giving a new driver the sensation of being in a fish bowl. Impact absorbing crash boxes in the front add to standard passive safety features.

Adding to its exotic design, the car has no real dashboard and no pedals taking up floor space inside. Nor does it have a conventional steering wheel. In the conventional rack-and-pinion steering system, when the driver turns the steering wheel a shaft connected to a pinion gear rotates, which moves a rack gear connected to the car's front wheels. In the Hy-Wire, what replaces the conventional steering wheel is a control unit called the **X-drive**.[1] The X-drive uses a concept that GM calls "drive-by-wire" which is similar but much less complicated than modern jet fighter controls, which are called "fly-by-wire." The interface between the X-drive and the car is solely by electrical wires. Unlike the controls of standard cars of today which require mechanical linkages, the X-drive uses only a single electrical cable connected between it and the skateboard. This is another reason why different car bodies can be easily connected to the skateboard. The X-drive allows the driver to steer, accelerate, and brake using hands only. The driver's feet are not used at all, except to get in and out of the vehicle.

While these features alone would make the Hy-Wire different from conventional cars, the defining characteristic of the Hy-Wire is its hydrogen PEM fuel cell stack. The fuel-cell stack, called the HydroGen3, is composed of 200 individual cells connected in series. It's about the size of a personal computer tower and provides 94 kW of continuous power and 129 kW of peak power. To cool the system, the stack powers a radiator system.

The Hy-Wire hydrogen fuel cell stack powers an electric motor that drives the wheels. The electric motor's controller converts 125-200 volts DC to 250-380 volts AC to feed the **three-phase electric motor.** The amount of voltage used depends on the speed. An onboard computer increases or decreases the power provided to the motor in response to the driver's speed commands. When the controller applies maximum power from the fuel-cell stack, the motor's shaft can spin at 12,000 rpm, delivering a torque of 159 pound-feet. At maximum power, a single-stage planetary gear with a ratio of 8.67:1, steps up the torque to apply a maximum of 1,375 pound-feet to each front wheel. This means that the Hy-Wire can easily travel 100 miles per hour on a level road. Smaller electric motors change the direction of the wheels based on the driver's steering commands, and electrically controlled **brake calipers** squeeze brake pads to bring the car to a stop. Though impressive, like most other fuel cell cars, the ones from GM nonetheless face several challenges if they are to become commercially viable.

Challenges for the Hy-Wire

The challenges whose solutions would have the greatest impact are

    Onboard hydrogen storage
    Durability
    Cost

GM has chosen to store hydrogen gas at 5,000 psi in three cylindrical tanks that weigh about 165 pounds total. The weight and the special carbon composite materials built into the tanks give them the high structural strength needed to contain hydrogen gas under such high pressure. The tanks hold enough hydrogen to travel approximately 90 miles, which isn't very much. To increase tank capacity, GM engineers are currently looking into ways to increase the pressure. The bigger challenge to date, however, is to improve durability. Whereas the ICE in cars can run for about 5000 hours or 150,000 miles, the PEM fuel cell in cars can currently run for only about 1,500 hours or 45,000 miles.[2] Until these challenges are met and the cost of the car is reduced, the fuel cell car itself is unlikely to be commercially viable.

19
Stationary Power Fuel Cells

While hydrogen conversion devices and applications in the transportation sector are not yet commercially available, hydrogen fuel cells in the stationary power sector are reaching the commercial market. One of the advantages fuel cells have in the stationary power sector is that there are more to choose from because they have fewer constraints on temperature.[1]

Along with the PEM fuel cell, four other primary types of fuel cells are being tested as sources of stationary power and as replacements for hydrocarbon-based electrical power: 1. Alkaline Fuel Cell (AFC), 2. Phosphoric Acid Fuel Cell (PAFC), 3. Molten Carbonate Fuel Cell (MCFC), and 4. Solid Oxide Fuel Cell (SOFC).[2] The characteristics of the five basic types of fuel cells are summarized in Table 2.

Table 2. Fuel Cell Characteristics[3]

	PEM	AFC	PAFC	MCFC	SOFC
Electrolyte	Perfluorosulfonic acid polymer	Aqueous Potassium Hydroxide	Phosphoric Acid	Molten Potassium Carbonate	Yttria-stabilized Zirconium Oxide
Operating Temperature	120-212°F	190-220°F	300-400°F	1112-1300°F	1200-1850°F
Anode Catalyst	Platinum	Nickel or Precious Metal	Platinum	Nickel / Chromium Oxide	Nickel / Yttria-stabilized Zirconia
Cathode Catalyst	Platinum	Platinum or Lithiated NiO (and Others)	Platinum	Nickel Oxide	Strontium doped Lanthanum Manganite
Heat Co-generation	Low Quality	None	Medium	High	High
Efficiency	25-55%	60-70%	30-40%	45-50%	35-50%

The most obvious difference between the fuel cells is their temperature range. The PEM fuel cell can operate at a relatively low temperature range of 120-212°F, while at the other extreme the SOFC requires temperatures at or above 1200°F. Why the temperature differences? Each type of fuel cell has its own minimum temperature requirement for the transfer of ions across the electrolyte.[4] In addition to the minimum operating temperature, the electrolyte also determines most of the other operating characteristics of a fuel cell.

The operation of the MCFC, shown in Figure 27, can serve as an example.

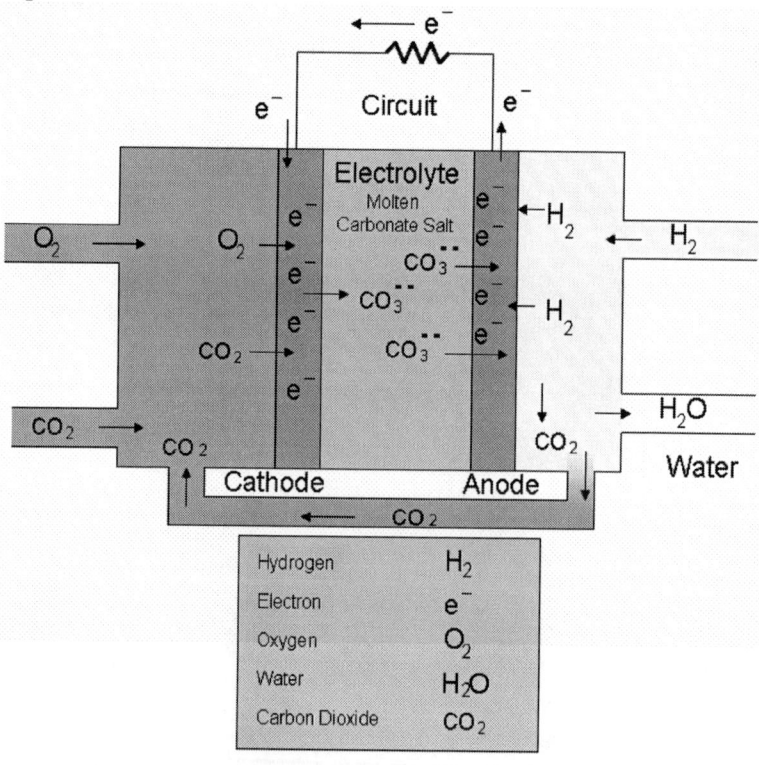

Figure 27. Molten Carbonate Fuel Cell (MCFC) [5]

On the anode side, hydrogen is introduced and reacts with carbonate ions from the electrolyte in the presence of the nickel chromium oxide catalyst of the anode. The products of the reaction are water, carbon dioxide, and electrons. The electrons travel to the cathode through the electric circuit, which is external to the fuel cell. The water is drained away. The carbonate ions (CO_3^-) in the electrolyte are replenished using carbon dioxide from the reaction at the anode. The carbon dioxide is recycled to the cathode side. At the cathode side, it interacts with the catalyst and with oxygen to produce carbonate ions, which enter the electrolyte. The MCFC operates at a temperature of around 1200°F, which is the temperature at which carbonate ions move freely through the electrolyte from the cathode to the anode.

Each of the fuel cell types has pros and cons. For example, the MCFC works well with cheaper catalysts like nickel and has a relatively high efficiency of 50%. Current MCFC demonstration units have produced up to 2 MW, but designs exist for units that can produce 50-100 MW. Also, the MCFC is more tolerant of impurities than is the PEM. The MCFC is less prone to carbon monoxide contamination of the catalysts in its electrodes than either the PEM or Alkaline Fuel Cell (AFC) because of the MCFC's higher operating temperature. Since the MCFC is tolerant of impurities such as carbon monoxide, it can use a fuel cell **reformer,** which is a device that extracts hydrogen from petroleum, natural gas, coal, or a variety of other hydrogen containing compounds.[6] Usually the reformer leaves some impurities in the extracted hydrogen, which is why not all fuel cells can use it.

On the downside, the liquid nature of the MCFC's electrolyte places greater constraints on the MCFC than on fuel cells that use solid electrolytes, like the solid oxide fuel cell (SOFC) or the PEM. It is not only more difficult to work with a liquid electrolyte than with a solid, but in the case of the MCFC it requires more ingenuity to get it to work. For example, even though carbon dioxide from the anode is recycled to the cathode, there may not be a sufficient supply of carbon dioxide from recycling for optimal MCFC performance. Still the advantages of the MCFC, especially with respect to fuel efficiency, make it a major contender when choosing between fuel cell technologies.

Stationary Power Hybrids

Like the transportation sector, the stationary power sector has also produced hybrids that may act as bridges to future energy technology. The fuel-cell/gas-turbine hybrid is one such example. In 2000, Southern California Edison together with the California Energy Commission, the Electric Power Research Institute (EPRI), the Department of Energy (DOE), and the South Coast Air Quality Management District (AQMD) became the first in the U.S. to create a prototype of a fuel-cell/gas-turbine hybrid, which they housed on the campus of UC, Irvine. The prototype cost about $16 million and generated 220 kW. This little power plant, which was the size of a small house trailer, combined a Siemens-Westinghouse solid oxide fuel cell (SOFC) and an Ingersoll Rand micro-turbine with some remarkable results.[7]

Breakthroughs in Hybrid Stationary Power

Everyone on the project recognized that the high operating temperature of the solid oxide fuel cell and its solid ceramic electrolyte made it suitable for integration with a gas turbine. What the Siemens-Westinghouse engineers didn't expect was how efficient and pollution free this configuration would turn out to be. The SOFC operating alone had an efficiency of 50%, and the Ingersoll Rand micro-turbine operating alone had an efficiency of 30%. As a hybrid, they found that the efficiency of the combined unit jumped to 60%. A key was the discovery that the power output of the SOFC could be increased by 10% if it was operating at a pressure of 3 atmospheres (44 psi). The improved efficiency affected the hybrid as a whole. In operation, the SOFC supplied approximately 180 kW while an additional 40 kW was obtained from the micro-turbines for a total of 220 kW. The hybrid's efficiency was enhanced further by the use of a recuperator to recover waste heat – an idea car manufacturers might look into.

Almost as remarkable as the efficiency and power was how little pollution a micro-turbine emits in this configuration in comparison to a micro-turbine alone. In the conventional non-hybrid micro-turbine, high temperature, high pressure gas rushes out of a combustor and pushes against the turbine blades, causing them to rotate. It is analogous once again to water rotating the turbine blades at Niagara Falls. This hot gas is obtained by burning a fuel – usually gasoline or natural gas – in a **combustor**, which is why gas turbines are sometimes referred to as combustion turbines. Yet not all of the fuel is fully combusted. Some of it ends up being expelled in the exhaust along with other pollutants. To increase efficiency, the turbines sometimes add the recuperator, which is a sheet-metal heat exchanger that recovers some of the heat from the exhaust stream and transfers it to the incoming air stream.

In the hybrid SOFC - gas turbine system with recuperators, hot steam from the SOFC, rather than combusted fuel, becomes the input to the turbine. There is no need for a combustor and this is the key to the lower emissions. The temperature of the SOFC exhaust is already between 1200-1850°F. Because it operates from an electrochemical process rather than combustion, the micro-turbine in the hybrid system emits virtually none of the air pollutants commonly released by conventional power plants. Levels of one pollutant it does emit, nitrogen oxide, are nearly 50 times less than levels emitted by today's typical natural gas turbine. This environmental friendliness means that the hybrid fuel cell - gas turbine system can be sited almost anywhere.

"This new technology has the potential to alter the landscape of tomorrow's power industry," Spencer Abraham, head of the Department of Energy at the time, said in a news release. He went on to say, "It offers a preview of the day when more of our electricity will be generated by super-clean, high-efficiency power units sited near the consumer. Distributed generation could play a key role in strengthening the security and reliability of our power supply, and fuel-cell/gas-turbine hybrids could help make distributed power a reality." [8]

Other Fuel Cells: Direct Methanol Fuel Cell

It may appear from what has already been said that fuel cells have to use hydrogen to operate. In fact, it is not the only option. The Jet Propulsion Laboratory in California and the University of Southern California discovered an alternative fuel cell that they call the direct methanol fuel cell. As the name implies, this fuel cell can produce electricity directly using methanol (CH_3OH). It doesn't require a reformer to produce hydrogen from methanol. In the direct methanol fuel cell, methanol reacts with oxygen from air to produce electricity with the same byproducts as the hydrogen fuel cell produces: water and carbon dioxide.

So far, the direct methanol fuel cell doesn't seem practical except for small applications such as computers. But that may change, given the pace of discovery and innovation in the last few years.

20

Hydrogen Production, Distribution, and Storage

Frustratingly, the great advances in hydrogen fuel cell technology and the progress made with the hydrogen internal combustion engine (ICE) have not been matched by equally impressive advances in the production, delivery, and storage of hydrogen. Arguably the weakest link in the hydrogen infrastructure today is hydrogen production.

Hydrogen Production

So what is the status of hydrogen production technology? Today, several techniques have been proposed to produce the hydrogen we would need. The leading hydrogen production techniques are (1) electrolysis (2) photo-electrochemical (PEC) reaction (3) fermentation (4) photo-biological reaction (5) reformation (6) thermo-chemical reaction and (7) coal gasification, which includes CO_2 sequestration. To these basic techniques can be added variations and combinations such as thermo-chemical gasification, which can also be used to produce oil from coal. Finally, there is the inverse process of making other biofuels such as methane and methanol from hydrogen. A survey of each basic technique can give a fairly good idea of where we're at with hydrogen production and where the possibilities for breakthroughs lie.

Electrolysis[1] – The advantage of electrolysis is that it derives hydrogen from water (H_2O), which is the most readily available and abundant source of hydrogen on the planet. Water is ubiquitous in modern societies, inexpensive, and safe to store, transport, and dispense. Electrolysis has a drawback, however. The electricity required to produce hydrogen from water has to come from somewhere. There are people who claim you can put water into the tank of your car or into some electronic device and use fuel cells to generate electricity from it.

That is a half truth. You have to generate electricity to extract from water the hydrogen that the fuel cell needs to generate electricity. Electrolysis and inverse electrolysis do not just cancel one another out. The energy input is greater than the output. Making this situation even more unacceptable, most electricity that could be used to produce hydrogen is currently generated using natural gas, gasoline, coal, and oil.

In order to use electrolysis effectively to produce hydrogen, it will be necessary, at a minimum, to find a renewable energy source from which to generate the electricity required for the electrolysis. Unfortunately, at present only about 3% of the electricity consumed in the United States is generated by energy sources such as biomass and corn, wind turbines, or solar energy while only about 7% comes from hydroelectric plants. This small percent may be compared with the 19-20% of electricity generated using nuclear plants and approximately 49% generated using coal.

Photo-electrochemical (PEC) Reaction. This is a method of producing hydrogen using solar energy. It is sometimes viewed as artificial photosynthesis. A PEC cell produces hydrogen by using a photovoltaic (PV) semiconductor electrode, a metal electrode, and an aqueous electrolyte. When light is incident to the PV semiconductor electrode, the electrode absorbs some of the light (photons) and generates electricity. The electricity then splits the water contained in the aqueous electrolyte into hydrogen and oxygen. Some advantages of PEC systems are: 1) no external electricity is required 2) it is potentially non-polluting[2] 3) it is scalable (can be made any size as needed) which makes it a candidate for either distributed or centralized production.

Up until the year 2000, the advantages of PEC cells had been offset by the danger of electrolyte leakage, but in May of 2000, Toshiba announced it had discovered a solid electrolyte for the PEC cell, eliminating the problem.[3] However, challenges remain. The solar to hydrogen conversion efficiency of PECs has been only 7-11%, which doesn't make them very cost effective given the quantities of hydrogen that would be needed. PEC hydrogen production systems are also limited because they need light to operate.

For PEC systems to be commercially viable, breakthroughs have been required to improve the solar to hydrogen conversion efficiency, semiconductor electrode durability, and PEC design so it can be more easily manufactured for a mass market.[4] In June 2006, one of the breakthroughs was achieved. Michael Grätzel and his colleagues at the

Ecole Polytechnique Federale de Lausanne in Switzerland announced they had developed a photoelectric device that increases the efficiency of splitting water into hydrogen and oxygen to 42%.[5] With this advance, the field of photo-electrochemical (PEC) hydrogen production is getting more exciting and promising. But costs remain a problem.

Fermentation and MFC. While running electricity through a hydrogen compound in mimicry of lightning might seem like the only possible method of producing hydrogen, it is not the only method derived from nature. Biological organisms discovered other methods millions of years ago. Fermentation is one. Fermentation has long been used to produce ethanol. It can also produce hydrogen.

More recently, the Pennsylvania team that developed the microbial fuel cell (MFC) announced a new discovery related to their fuel cell. It is a new process for producing hydrogen. They found that when a small amount of electricity is applied at the cathode, the MFC bacteria will produce over four times more hydrogen than is generated from fermentation alone.

The importance of discovering an electrically enhanced MFC can be seen, according to the researchers, by a comparison. For conventional electrolysis, the electricity required to produce 1 cubic meter (m^3) of hydrogen is 5 kWh.[6] This amount is more than eight times the 0.6 kWh input required for the electrically enhanced MFC to produce the same amount of hydrogen. Viewed another way, using the MFC to produce 1 mole of hydrogen requires the electrical energy equivalent of 0.2 moles of hydrogen. This is an EPR of 5:1.

In contrast, using conventional electrolysis to produce 1 mole of hydrogen requires, according to the Pennsylvania team, the electrical energy equivalent of 1.6 moles of hydrogen. This is an EPR of 0.6:1,[7] a net energy loss. Not everyone would agree with this degree of energy loss, however.[8] But there is a net energy loss in any case. If we use electricity generated from conventional hydrocarbons to produce hydrogen, we are frittering away a resource that can not be renewed. But with an MFC, we get a large net energy gain.

The Pennsylvania researchers concluded, "It should be possible for the first time to produce large yields of hydrogen from domestic wastewater using this approach. Combining hydrogen production and wastewater treatment should result in a more economical venue for hydrogen generation as the infrastructure needed for wastewater treatment can be used to effectively subsidize the cost of the hydrogen generation."[9] The researchers call their enhanced MFC the Bio-Electrochemically-Assisted Microbial Reactor or BEAMR.[10]

It remains to be seen whether the predictions based on their experimental research will pan out, but the idea of creating a microbial reactor that serves multiple purposes is a powerful one. Multi-purpose alternative energy techniques or devices could make the difference for alternative energy in the marketplace by making them profitable.

Photo-biological Reaction. The BEAMR is not the only discovery in the field of biology that could further the quest to produce hydrogen. It is a remarkable fact that the same primitive micro-organisms that first supplied Earth's atmosphere with oxygen are capable of switching from oxygen production to hydrogen production under certain conditions.[11] We owe our existence to these ancient micro-organisms and in the future we may owe our continued high standard of living to them as well, at least in part. Blue-green algae, green algae, and photosynthetic bacteria are all being studied for their capability to use sunlight to produce hydrogen, chiefly from water. The leading contender to produce hydrogen in large quantities is a species of green algae called *Chlamydomonas reinhardtii*.

C. reinhardtii were first cultured for hydrogen at the University of California, Berkeley in 1998, where it was found that they could produce hydrogen when switched to a sulfur-deficient, anaerobic environment that may suggest their original primitive environment. The hydrogen was produced for only a limited period of time, after which *C. reinhardtii* had to be switched back to its normal environment (with sulfur and oxygen) in order to revitalize and be ready to continue hydrogen production.

Since their initial discovery, the researchers at Berkeley have managed to produce hydrogen continuously using a series of two flow reactors that provide the proper environment. This was a critical process advancement that has made the Berkeley research a center of attention.

The reason photo-biological systems and fermentation systems like BEAMR are potentially so important are their considerable advantages. Like photo-electrochemical devices, they require no external electricity and are non-polluting.[12] The photo-biological reaction doesn't require any special water purification system either. Hydrogen production using this technique may be either distributed or centralized. All that is needed is the proper habitat for the algae, which can vary in size. The press has dubbed this environment "slime ponds." In actuality, the environment consists of the photo-bioreactors, a means of hydrogen extraction and storage, and sunlight.[13]

So what have been the challenges? Because photo-biological processes depend on light, they do not operate at night and can lose efficiency on cloudy days or at greater pond depths. Even when there is enough light, photo-biological hydrogen production has been very inefficient. The microorganisms normally have a solar-to-hydrogen conversion efficiency of only 0.5%, which is abysmally low for commercial use.

DOE's funding[14] of research into photo-biological hydrogen production has been aimed specifically at overcoming the limitation on efficiency. It set a goal of 5% efficiency by 2015. That is still low. To be at all commercially viable, the efficiency must be at least 10%.

The DOE identified two basic areas in which breakthroughs must be made in order to increase efficiency and reduce costs: genetics and photo-bioreactor engineering. The genetic breakthroughs aim at increasing the green algae's capability to produce hydrogen by removing or reducing inhibitors. For instance, researchers would like to reduce the green algae's ability to co-produce oxygen, which inhibits the algae's secret to producing hydrogen, a hydrogen catalytic enzyme. *C. reinhardtii* not only harbors this catalytic enzyme called hydrogenase, it manufactures the substance itself. Hydrogenase acts something like catalysts in electrolysis.

Researchers are also taking aim at the size of the chlorophyll antenna that the algae use for direct photo-biological hydrogen generation. Under bright sunlight, the chlorophyll antenna absorbs more photons than can be used for photosynthetic electron transport, resulting in heat dissipation and a loss of up to 80% of the absorbed light. By inefficiently absorbing so much light, the algae higher up in the bioreactor prevent some photons from reaching algae at lower levels, reducing overall production.

Photo-biological hydrogen production today is still at the basic R&D stage. Nonetheless, photobiology has attracted commercial interest. One company, Melis Energy, was set up with the intention of eventually producing commercial hydrogen using green algae.

In 2006, an associate of Melis from Berkeley suggested that Tasios Melis, the founder of Melis Energy and one of the key Berkeley researchers investigating *C. reinhardtii*, had surpassed DOE expectations for 2015 and reached an efficiency of 10%.[15]

Reformation. While photo-biological systems are among the newest systems developed, reformation is the most commercially mature. Reformation technology adds thermal reaction to the catalytic conversion mechanism somewhat similar to those used by green algae.

There are three fundamental types of reformers currently in use: steam methanol reformers (SMR), auto-thermal reformers (ATR), and partial oxidation (POX) reformers. The differences between the three aren't great. Steam reformers use steam; partial oxidation units use oxygen; and auto-thermal reformers use both steam and oxygen. The operating temperatures vary. All require catalysts and intense heat. Steam reformation is the most ubiquitous. Around 95% of all hydrogen produced today is produced using steam reformation. Unfortunately, steam reformers require hydrocarbons and produce pollution.

Steam methanol reformers produce hydrogen from methanol (CH_3OH). Methanol is toxic. Ethanol, which is not toxic, can also be used. In the production of hydrogen from methanol, methanol's carbon atom usually combines with oxygen to produce carbon monoxide (CO) or carbon dioxide (CO_2). SMR engineers try to reduce the amount of CO they produce by adding water to their reformer. Water added to liquid methanol is vaporized, and the mixture channeled into a chamber where the mixture encounters a catalyst that splits it into CO_2, hydrogen, and CO. This process is called the **water-gas shift (WGS) reaction**.[16] Without water, the amount of CO produced would be much greater. But pollution still remains a problem.

Natural gas steam reformation is a closely related process that obtains hydrogen gas from methane (CH_4). The initial decomposition requires a catalyst to split methane and water into hydrogen gas and CO gas.[17] These gases are burned in the presence of another catalyst, using a little air to supply oxygen so that most of the CO is converted into CO_2.

The reason such care is taken to convert CO into CO_2 is not because CO is very poisonous to the human system but because it contaminates fuel cells. Direct use of the hydrogen from natural gas steam reformers by fuel cells at the same plant would be problematic without CO to CO_2 conversion.

Because natural gas can contain other pollutants such as sulfur, additional catalytic processes are required. They are usually introduced before the start of steam reformation in order to segregate the pollutants from the mixture and transform them into more harmless substances. Steam reformers operate at 1292-1697°F (700-925°C)[18] and have an excellent 65-75% efficiency rating, which is the percentage of hydrogen in their feedstock gases that they are able to extract. The main drawback to using reformation, beside pollution, is the use of fossil fuels. Steam reformation, if dependent on fossil fuels, is obviously not the way to reduce our use of fossil fuels.

Thermal Chemical Reaction. Thermal chemical reaction is a means of producing hydrogen that comes from the oil refinement process. This technique, also called thermal cracking, has already been described. Briefly, thermal cracking uses heating and then cooling to separate out different compounds from oil. The same basic technique of using temperature changes to separate elements out of compounds can be used to separate hydrogen from substances in which it is found.

Coal Gasification With CO_2 Sequestration. This process has been used to produce electricity without combustion. It can also be used to produce hydrogen from coal. There are a number of variations on how this might be done but a typical process has three stages: gasification (also known as hydro-gasification), carbonation, and calcination.

During the first stage, **gasification**, coal is subjected to heat, water, and a catalyst to convert it to a gas and to extract methane (CH_4). The output of the gasification stage is methane and water, in the form of steam.[19] Ash and other impurities are filtered out.

During the second stage, **carbonation**, methane reacts with steam (water) and calcium oxide (CaO) in the presence of another catalyst to produce hydrogen and calcium carbonate ($CaCO_3$).[20] Purifiers and filters scrub the hydrogen to make it pure. Some hydrogen is also fed back to the gasification stage, reducing the overall hydrogen production efficiency.[21]

During the third stage, **calcination**, the calcium carbonate is transformed into calcium oxide and carbon dioxide. The calcium oxide is fed back to the carbonation process, while the carbon dioxide is sequestered.

The sequestration process is supposed to permanently sequester (contain) the CO_2 at a selected site. The most likely candidates to become CO_2 containers are geologic formations such as a depleted gas reservoir, a deep saline aquifer, or a basalt formation, all of which are common in the United States.

HYDROGEN PRODUCTION, DISTRIBUTION, AND STORAGE   197

Figure 28 provides an example of how hydrogen can be extracted from coal using the three stages I've described. In the example, the hydrogen that is produced is used to run a fuel cell.

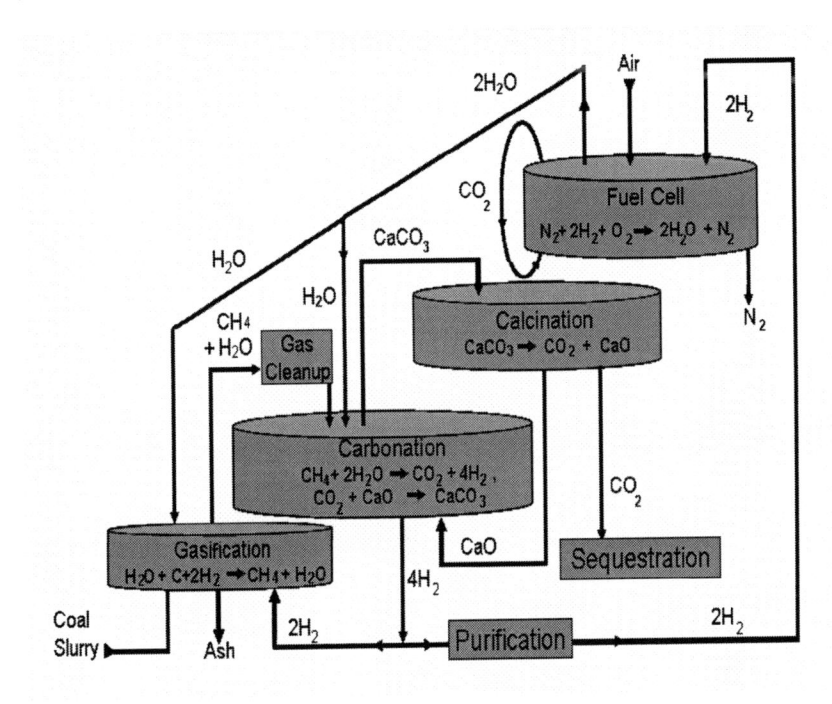

Figure 28. Hydrogen from Coal Gasification with Sequestration[22]

The DOE has set specific goals for the coal gasification with CO_2 sequestration process. It expects near-zero emissions by 2020; electricity with less than a 10% increase in cost compared to non-sequestered coal use; and it expects coal to become a major source of hydrogen by 2020.

One question that might be asked is, why use coal to obtain hydrogen given that coal is the dirtiest of the hydrocarbons? An answer is that coal gasification is both more efficient and less polluting than coal combustion.

Perhaps an even more searching question might be, why not stop at the first stage, the production of methane? Producing synthetic natural gas (methane) from coal would be a great boost to our stationary power economy. Synthetic natural gas is much cleaner than oil. It would come from domestic sources. Using coal gasification to produce synthetic natural gas seems to be a better use of coal and to be more cost effective.

198   POWER TO CHANGE THE WORLD

The main worry I'd have is how much methane might escape into the atmosphere, given that methane is more than 21 times more potent a greenhouse gas than CO_2.

Using Sunlight, Air and Hydrogen to Produce Renewable Fuels

While the end result of hydrogen production is usually thought to be hydrogen, it may make more sense to use hydrogen to produce other, heavier fuels. One example of how this could be done involves the setup in Figure 29. The process is straightforward - in theory.[23]

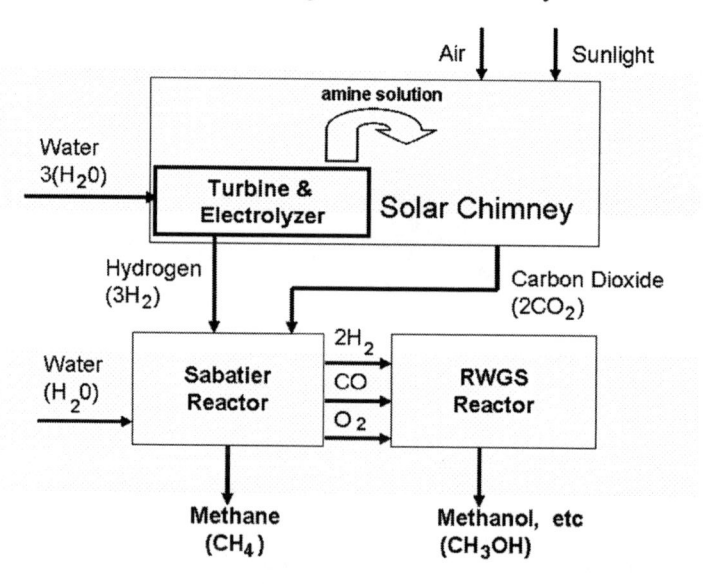

Figure 29. Block Diagram of Solar Tower[24]

Here is how the process might work. The solar tower in Figure 29 has a chimney-like structure in which air is heated by the sun. Since hot air rises, it moves up the chimney so long as the sun heats the air at the base. The upward flow of air is used to turn turbines which generate electricity. Some of the electricity is used to run a pump which pushes an amine solution to the top of the tower from which it is allowed to fall down into the upward flow of air inside the tower. As it falls, the amine solution captures CO_2 from the air. The amine solution with the captured CO_2 is collected at the bottom of the tower and heated to release CO_2

into a container. The rest of the electricity generated by the turbines is used to produce hydrogen from the electrolysis of water.[25]

Once sufficient amounts of hydrogen and carbon dioxide have been accumulated, the gases are piped from their respective containers into a Sabatier reactor, which is a steel pipe filled with a ruthenium alumina catalyst. A set of needle valves and a meter control the flow of gases from their respective containers into the reactor. When hydrogen and carbon dioxide flow into the reactor, they react in the presence of the catalyst to form methane (CH_4).[26]

What is required to produce the Sabatier reaction is H_2 from water, CO_2, the catalyst, a pressure of 11.6 psi, and a temperature of 482°F. Initially the Sabatier reactor needs to be heated to the required temperature by a set of nichrome heaters or their equivalent. However, once the reactor is in full swing, the Sabatier reaction generates its own heat via exothermic (heat liberating) chemical reactions.[27] A reverse water gas shift (RWGS) reactor can be added to the Sabatier reactor to produce additional fuels.[28] The RWGS reactor (provided with hydrogen, carbon monoxide, and oxygen from the Sabatier reactor) can use another catalyst, heat, and pressure[29] to produce methanol (CH_3OH),[30] ethylene (C_2H_4), or ethanol (C_2H_5OH). Only the main outputs of the process are shown in the diagram.

Why use hydrogen to produce renewable fuels like methanol instead of using hydrogen itself for fuel? There are two basic arguments for this course of action. First, because they are liquid at room temperature, renewable fuels like methanol and ethylene fit better within our current energy infrastructure than hydrogen does. They don't need a new infrastructure. Second, they can be produced using CO_2 drawn from the air[31] rather than from hydrocarbons so that no CO_2 is added to the atmosphere. The challenge is to find a way to make these fuels efficiently in sufficient quantities.

From Mars to Earth

Chemists have known about the RWGS reaction and the Sabatier reaction since at least the beginning of the nineteenth century. Yet the idea of using these reactions with solar energy to create fuels is very new. Dr. Robert Zubrin, an aeronautical engineer who worked for NASA, suggested the fuel-related application. His ideas appeared in a paper on NASA's manned missions to Mars and was based on his work in the 1990s. Zubrin proposed that rocket fuel could be manufactured on Mars for a return trip to Earth using hydrogen, the Sabatier reaction, and Martian CO_2, which is nearly 95% of the atmosphere.[32] Dr. Zubrin's ideas seem to have spurred research into how to use CO_2 and hydrogen to make renewable hydrocarbon fuels for earthly uses. Creating renewable fuel from the CO_2 in the air and hydrogen in water is not the only option. Fuel could be produced using captured CO_2 and hydrogen emitted from combusting or gasifying coal. Capturing CO_2 from coal emissions to make ethanol or methanol may be preferable to sequestering it; but studies will be required to make that determination.

How Much Hydrogen is Needed?

The U.S. government has also described the likely sources from which to produce hydrogen. They include the nuclear option (Figure 30).

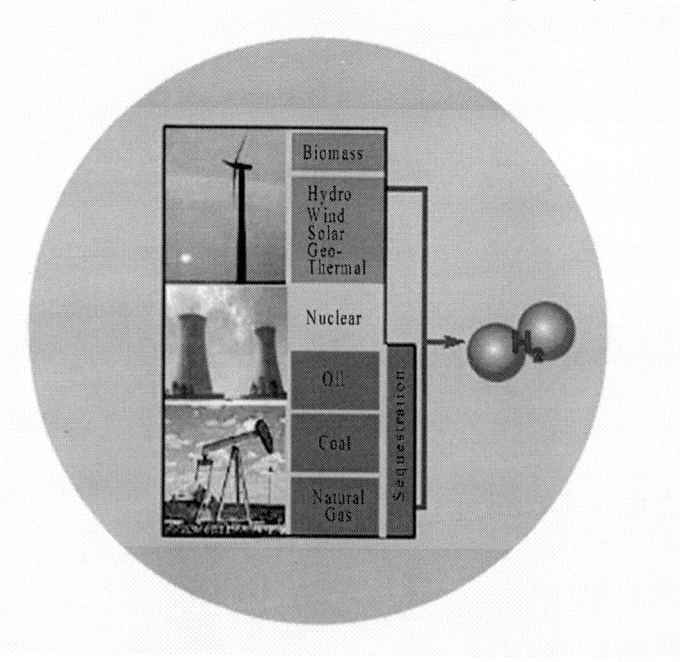

Figure 30. Hydrogen Production Techniques and Sources Summarized [33]

The government in its *Hydrogen Posture Plan* has also come up with a rough estimate of how much hydrogen would be needed, though the estimate is far from comprehensive. Their researchers estimate that 40 million tons of hydrogen would have to be produced annually to fuel 100 million fuel-cell powered cars and provide electricity for about 25 million homes.

Here is the U.S. Government energy researchers' breakdown of where some of the hydrogen could come from in the future:[34]

100,000 home electrolyzers-	4 million tons
15,000 small reformers in fuel stations-	8 " "
30 coal/biomass gasification plants-	8 " "
10 nuclear plants splitting water-	4 " "
7 large oil & gas SMR / gasification refineries-	16 " "

At present the U.S produces 9 million tons of hydrogen a year, according to the *Hydrogen Posture Plan*, which claims that this would be enough to fuel 20-30 million hydrogen cars or enough to power 5-8 million homes. However, nearly all of this hydrogen is derived from hydrocarbons or nuclear reactions. And of the 9 million tons of hydrogen produced per year, most is used in the refinement of petroleum, in the treatment of metals, and in the hydrogenization of fats.[35] Only a tiny fraction of the hydrogen produced is actually used as fuel. Among government agencies, the most notable user of hydrogen for fuel is the National Aeronautics and Space Administration (NASA).

In summary, multiple avenues for producing hydrogen exist. Figure 30 enumerates the main ones. But current production capability is low. Failure to increase quantities dramatically will delay the widespread use of hydrogen. Furthermore, even if production is stepped up so that sufficient quantities are produced, we will also need to improve hydrogen storage and handling, and find effective delivery systems.

Delivery Systems

The government has proposed at least five different delivery systems for hydrogen. Most require significant changes in our energy distribution infrastructure. The delivery systems are pipeline, on-site production and storage, packaged on-site production and distribution, liquid hydrogen transport by truck, and hydrogen transport by tube trailer.[36]

While these delivery systems may seem conventional, they each pose significant engineering and economic challenges. The key obstacle is that all the methods considered appear extremely expensive to implement. Few pipelines exist for hydrogen transport. Hydrogen as a light gas can not utilize pipelines that exist for conventional fuels. Liquefying hydrogen for transport by truck is costly and requires significant expenditure of energy. This expenditure further reduces the EPR, the ratio of energy output to input, for hydrogen. Carrying gaseous hydrogen by tube trailer in high-pressure cylinders is feasible, but much less hydrogen can be carried in a truck when it's a gas than when it's a liquid, so many more trucks would be needed. Whether liquefied or gaseous, hydrogen transported by trucks could be dangerous.

Breakthroughs in storage of hydrogen under low pressure, and breakthroughs in the efficiency of electrolysis, photo-electrochemical reaction, microbial fuel cells, and other techniques could make on-site production of hydrogen from water viable. Water is not only ubiquitous, it is also a nonflammable, clean, abundant source of hydrogen. It doesn't contain polluting carbons. And water is a major source of oxygen, making it the perfect feedstock for fuel cells that run on hydrogen and oxygen. But production and storage of hydrogen at or near the site where it is used is not favored in federal government plans.

Onboard Storage and Limited Hydrogen Availability

The range of a hydrogen car is limited by the amount of hydrogen it can store on board as well as the amount of hydrogen available. To address these problems, BMW has resorted to a bridging hybrid solution that allows their hydrogen cars to also run on gasoline. To increase efficiency, BMW has also added an electric motor powered by a 135 kW hydrogen fuel cell stack. All these additions make the BMW hydrogen car the "everything car." They claim the driving range of their hybrid is about 600 miles.

Car companies are not the only companies trying to overcome the problems of limited hydrogen fuel resources and limited storage capacity.[37] Texaco joined with Ovonics Energy Conversion Devices, Inc to form a company called Texaco Ovonics Hydrogen Systems LLC (TOHS). Their mission was to solve the onboard storage capacity problem in a dramatic way. As they put it:

> "We believe that one of the key enabling technologies for implementing the hydrogen economy now is our metal hydride storage solution ... Our storage system could be used today in cars and trucks

that operate with either internal combustion engines or fuel cells, it could also be used as an alternative to batteries for power back-up systems, as a fuel alternative for on-site generation, and it could be installed in just about any end-use application that today uses traditional hydrocarbon fuels."[38]

TOHS' solid hydrogen storage system consists of a tank containing metal hydride that absorbs hydrogen like a sponge when pressurized and releases the hydrogen when depressurized.

The advantages of this metal hydride storage system over an ordinary pressurized tank are impressive, if Ovonics figures are correct. According to Ovonics, containers with metal hydrides are able to store 3 kilograms of hydrogen at 2,000 psi, while compressed hydrogen containers can store only 0.88 kilograms at 3,600 psi;[39] that is, metal hydrides not only store more than 3 times as much hydrogen, they do so at a little over half the pressure. According to Ovonics, the metal hydride system has gone through hundreds of refilling test cycles with no performance degradation, is fully scalable, and can be adapted for use in a variety of mobile delivery tanks as well as in larger stationary storage systems.

In the *National Hydrogen Energy Roadmap,* the U.S. Department of Energy describes canisters of metal hydride as one of three leading storage mechanisms. The other two are compressed gas cylinders and liquid hydrogen containers.[40] While storage remains a weak link in the government's proposed hydrogen energy infrastructure, TOHS results suggest that the storage problem can be solved. Metal hydrides, if their potential is realized, are particularly desirable because they render hydrogen less flammable and make it more portable and distributable.

Codes and Standards

There is one other absolutely essential component of a hydrogen energy infrastructure: codes and standards for producing, transporting, storing, handling, and using hydrogen. Some codes and standards already exist for handling hydrogen, thanks to the pre-existing hydrogen industry. And a number of truck industry regulations exist regulating the transport of combustible gases and materials. In addition, a coordinating committee responsible for hydrogen codes and standards has been set up. And an international code council has been created that focuses on establishing safe handling practices, eliminating barriers to international trade in hydrogen, and developing criteria for quality assurance and methods of testing.[41] Most of the codes and standards are devoted to common sense precautions.

21

Technological Readiness Assessment

A careful assessment of current alternative technology suggests that we do have the wherewithal to reduce our dependency on foreign oil and decrease the environmental impact of hydrocarbons significantly. But we face some daunting technological challenges across the alternative energy field. The five most important are:

Cost
Quantities/Scalability
Limitations on Use
Durability/Longevity
Efficiency

Because of the challenges we face, we will have to rely for some time on a combination of near-term solutions to our energy problems rather than on one all encompassing solution. Advocates of single-solution "economies" such as a "hydrogen economy," or a "methanol economy," or a "solar economy" are not being realistic if they suggest a single-solution economic model is a short-term possibility.

Because the components of a new energy infrastructure differ in maturity, the technology needs to be examined and prioritized not only according to near-term and far-term potential but in terms of how it fits within the existing infrastructure. Fortunately, most of the near-term options are what I've called bridging technologies, technologies like the hybrid car that can exploit the existing energy infrastructure and provide us time to evolve and mature an alternative.

Though no single solution is on the horizon, that doesn't mean we should not explore solar power or hydrogen as part of a package of clean energy technologies. We not only should, we have to. So long as we fail to make clean energy a true alternative, we will be forced to rely on oil, coal, and nuclear power, with the risks they entail.

22

The Lack of Visible Progress

Though technological solutions to our energy problems are crucial, the challenges to transforming America's energy infrastructure aren't just technological. They are also political, economic, and social. All the technological prowess we possess will be of little consequence if we don't have the understanding, foresight, and will to act on our own behalf and on that of our nation. We took up the challenge of Sputnik in the 1950s, ending it with our triumphant landing on the moon. But we failed to sustain efforts to free ourselves from foreign oil despite the obvious threat the 1973 embargo revealed. Instead we slid back into even greater dependency. Why? To solve the problem we have with energy requires the broader insight into what holds us back, and how we can move forward.

Finally a Federal Plan

If money spent on alternative energy research was the principal measure of success, the United States would be leading the world. The Department of Energy (DOE) funds important ongoing projects in all the alternative energy technologies I've mentioned and more. Moreover, soon after the events of September 2001, the DOE did something that hadn't been done before. In November 2001 at the request of the Bush administration, the DOE came up with a plan whose goal was to regain our energy independence. This plan was set forth in *A National Vision of America's Transition to a Hydrogen Economy - To 2030 and Beyond*.[1] The DOE then published the *National Hydrogen Energy Roadmap*[2] in November 2002 and the *Hydrogen Posture Plan*[3] in February 2004. Both subsequent documents expanded on and added details to the *National Vision* plan. As almost all their titles suggest, all these plans focus on only one alternative – hydrogen. In determining what would be

THE LACK OF VISIBLE PROGRESS 207

in the *National Vision,* the DOE called upon "The efforts of ... 53 business executives, federal and state energy policy officials, and leaders of universities, environmental organizations, and National Laboratories..."[4]

The assembled leaders demonstrated that they clearly understood not only what is at stake ("the life's blood of the nation"), but also the difficulties we face in bringing about a hydrogen economy. They acknowledged, for instance, the problems of developing the new infrastructure,

> "....Even when hydrogen utilization devices are ready for broad market applications, if consumers do not have convenient access to hydrogen as they have with gasoline, electricity, or natural gas today, then the public will not accept hydrogen as 'America's clean energy choice.'"[5]

They recognized that the costs would be formidable for maintaining our current infrastructure while trying to build up a parallel immature alternative. "America's energy infrastructure is aging and in need of significant upgrades, overhauls, and replacements over the next several decades."[6] Despite these concerns, they lent their names to a plan that calls for:

- Accelerating the pace of research into hydrogen
- Assisting industry in developing and demonstrating fuel cells, and in creating a hydrogen infrastructure
- Assisting businesses in lowering the costs of hydrogen production, storage, and distribution
- Promoting the use of renewable energy sources of hydrogen
- Developing appropriate safety codes and equipment standards
- Improving public awareness of hydrogen

With their documented plan, the DOE proposed a timetable for moving from where we were in 2001 to where the government thinks we should be by 2035.[7] It calls for hydrogen power and transport systems to be commercially available between 2015 and 2035.

Under this plan, transition to a hydrogen economy will evolve in four phases with the final phase occurring sometime between 2030 and 2040. During phase one the DOE envisions the federal government playing a key role in funding research from which markets can develop. During the second phase, it expects the federal government to become an early technology adopter and to enact policies that nurture further development of hydrogen. During phases three and four, it expects a burgeoning hydrogen energy industry to take over the dominant role of establishing and spreading the necessary infrastructure.

To achieve its goals, the DOE plan has set a number of key milestones for individuals, groups, companies, and institutions (including universities) that want to receive funding. The major goals are to (1) lower the cost of producing and delivering hydrogen (2) provide more compact, light weight, lower cost, safe, and efficient storage systems (3) find low cost materials for advanced conversion technologies, especially for fuel cells (4) provide more effective and lower cost carbon capture and sequestration processes, and (5) provide designs and materials that maximize the safety of hydrogen use.[8] To strengthen its internal coordination, the DOE called for the establishment of a working group that would oversee its transitional milestones.[9]

Unfortunately, we have made little visible progress since the *National Vision* was initially published. Excluding nuclear power and hydroelectric power, we still obtain barely 3% of our energy from hydrogen and renewable alternative sources while we have become even more dangerously reliant upon imported oil. By these simple measures, the plan is failing.

It would be understandable if the public believed that the United States has not spent much at all on alternative energy since 1970. But this perception would not be correct. Without the DOE's investment of billions of dollars over the last thirty years, our energy prospects today would be much bleaker than they are. So while it is easy to blame the federal government for not doing anything about our continued dependency on hydrocarbons, it might be more productive to ask why there has been so little apparent progress, despite a lot of thought and effort.

23

What Holds Us Back

One reason we have made so little progress is a corrosive lack of consensus about what to do, despite the appearance of a consensus in the federal government's hydrogen plans. Not only has there been little consensus on what to do, there has been strong resistance to virtually every alternative proposed. Yet, while some are bogus and need to be addressed to prevent them from gaining greater currency, many of the arguments against the alternatives to our present hydrocarbon energy infrastructure have merit and must be taken into account. If we fail to take them into account, we are less likely to find a path forward. Without a firm sense of direction, it is no wonder we are in trouble.

Alternatives to our current hydrocarbon-based energy infrastructure have been critiqued from at least three very different vantages: those who believe there is no problem, those who believe there is a problem but that it can't be solved using alternative energy, and those who believe there is a problem but either it is too late to solve it or attempts at solving the problem put an emphasis on the wrong type of alternative energy. It is not unusual to find more than one of these points of view put forward in the same article or book.

What Problem?

The most divisive of the arguments that I've found are those involving hydrocarbon emissions, global warming, and toxic waste. The positions on global warming in particular spur heated exchanges and foster intransigence. Typical is this denial related to global warming,

> "The atmospheric data simply do not support the elaborate computer-driven climate models that are being cited by the United Nations and other promoters of the [Kyoto] Accord as 'proof' of a major future warming."[1]

The advocates of this position argue that there is no atmospheric evidence for major global warming. They don't specify what they mean by "major" but suggest we have little to worry about.

> "When dinosaurs roamed the tropical Dakotas, we didn't worry about SUVs polluting the air, because human activity didn't exist. But we had a very warm globe! So called global warming is not caused by man made greenhouse gases. Real scientists say, 'No'."[2]

In fact, real scientists agree that our globe is warming but differ on how much of a threat we face and whether it is manmade or natural or both. The public is left with vehement, contradictory arguments and accusations. The global warming issue is so emotionally and politically charged that it could polarize us, obstructing progress. Each side perceives a hidden agenda in the other, and many on both sides in this debate accuse their opponents of profiting economically from their position.

Some of those who believe that the global warming threat is overblown argue that leading promoters of it have a sinister desire to destroy capitalism for political reasons and personal profit. The sinister plan involves forcing the U.S. to reduce carbon emissions too quickly, which could ruin the economy. It is alleged they would also make the U.S financially liable for any extreme weather event in any country anywhere in the world on the grounds that the U.S. bears the main responsibility for the effects of global warming. American lawyers representing aggrieved weather damaged countries stand to earn a great deal of money, according to this argument, if claims are filed in certain U.S. courts.

A contrary claim, which is equally emotionally charged, is that wealthy American oil companies and car manufacturers are hindering solutions to a real problem in order to maintain their own short-sighted self interest and economic gain.

Consequently, arguing for change solely on the basis of global warming must run into problems. However, perceptions that something must be done are increasing momentum for change and could be strengthened by adding the argument for energy independence from hostile oil producing countries, which is much harder to refute.

Those who believe that "global warming is overblown" also tend to believe that "toxic pollutants aren't a worry" and that trying to deal with pollution is a waste of money and time. Here is an example:

> "Americans continue to pay a heavy price for their irrational fear of chemicals. Billions of dollars are being wasted on attempts to reduce toxics and other emissions to levels far below levels shown to have any negative effect on human health or wildlife."[3]

This claim is contradicted by a significant number of reputable studies that the pollution problem is very real, but most of the studies are local. We vitally need a more comprehensive understanding of pollution and its effects across the globe; a study JPL/NASA might become engaged in. If done, the results need to be widely publicized.

Then there are denials that oil is becoming scarcer. Typical is this rosy assessment,

> "Given the fact that the Earth shows no signs of running out of oil in the near or even far future, the notion of spending billions to replace it seems odd at best, foolish at worst. The Earth's reserves of oil have been consistently underestimated for decades since oil was first discovered."[4]

This assessment, whatever its other merits, makes no distinction between oil that is easy to get out of the ground and oil that isn't, or between domestic supplies of cheap oil and foreign supplies. These are fundamentally important distinctions with huge implications that must be pointed out. At the same time, we need to better assess the amount of oil left unexploited because of expense and limits in technology.

No Exit

In stark contrast to the sanguine assessment of those who see "no problem," some believe the problem is not solvable using alternatives and can only be mitigated by a massive scale down of life style. In an article entitled "The Long Emergency" James Kunstler writes,

> "Most immediately we face the end of the cheap-fossil-fuel era. It is no exaggeration to state that reliable supplies of cheap oil and natural gas underlie everything we identify as the necessities of modern life..."[5]

Having taken this position, he then dismisses one by one the alternatives to cheap fossil fuel as unworkable. He sees the hydrogen economy as "a particularly cruel hoax" and nuclear energy as a dim prospect no one should want. He writes that solar-electric systems and wind turbines "...face not only the enormous problem of scale but the fact that the components require substantial amounts of energy to manufacture and the probability that they can't be manufactured at all without the underlying support platform of a fossil-fuel economy." And he dismisses biomass with the observation, "Virtually all 'biomass' schemes for using plants to create liquid fuels cannot be scaled up to even a fraction of the level at which things are currently run."

So what are we left with? Kunstler's conclusion was, "No combination of alternative fuels will allow us to run American life the way we have been used to running it, or even a substantial fraction of it." As a result he says, "The circumstances of the Long Emergency will require us to downscale and re-scale virtually everything we do and how we do it..."[6] Nor does Kunstler see this decline in life-style as a gradual adjustment. He foretells a period of violence and disruption in which anger boils over, and in some parts of the country "substantial levels of violence ... collide with the delusions of Pentecostal Christian extremism." Kunstler's views were expanded in a book entitled *The Long Emergency: Surviving the End of the Oil, Climate Change, and Other Converging Catastrophes of the Twenty-First Century.*

In response to Kunstler's conclusions, I can say that fortunately his arguments are flawed. So is his conclusion that we are essentially helpless. For instance, his argument that solar and wind turbine components can't be manufactured at all without the underlying support platform of a rapidly depleting oil economy, overlooks the large coal reserves we possess.

Coal rather than oil can be used to supply power to manufacture the components of the alternative energy infrastructure we need. But we have to consciously choose what we are going to do with what we have. Will we continue to combust coal to generate electricity or will we choose to use cleaner gasified coal for electricity, which reduces pollution levels? Will we use coal to manufacture SUVs with poor fuel efficiency or will we use coal for manufacturing cars that can get 50-100 mpg? Will we continue to use coal in the manufacture of appliances that consume huge quantities of other fossil fuels or will we use coal to make appliances that use up to 75% less electricity? How we use fossil fuels is as important as the fact that we are compelled to continue using them in significant quantities in the manufacturing sector.

Kunstler can also be challenged regarding his claim that the energy expenditure required to create alternatives is not cost effective. A number of studies indicate that the energy required to produce solar panels and wind turbines is recouped within a few years.[7] The working life of a wind turbine is about 25 years. Wind turbines take six months of operation in a reasonably good wind area to recover the equivalent of the energy that went into their manufacture and construction, which is consistent with their phenomenal 50:1 EPR. Solar panels take 3-4 years of operation to recoup the equivalent of the energy that went into their manufacture,[8] which is consistent with their EPR of 7:1, and they last around 20 years.[9]

Something else that Kunstler doesn't take into account is that once up and operating, PV solar panels[10] and wind turbines emit zero air pollution.[11] By comparison, a typical conventional 500 MW coal power plant outputs more than 800 tons of pollutants per day.[12] Table 3 has a partial breakdown.

Table 3. Typical Daily Emissions for 500 and 200 MW Coal Plants[13]

Emissions	500 MW	200 MW	Units
Sulfur dioxide	27.4	11.0	Tons
Nitrogen oxide	27.9	11.2	Tons
Carbon dioxide	0.010	0.004	Tons
Small particles	1.37	0.55	Tons
Unburned hydrocarbons	0.603	0.241	Tons
Carbon monoxide	1.98	0.79	Tons
Ash	342.5	137.0	Tons
Coal Sludge	528.8	211.5	Tons
Arsenic	0.616	0.247	Pounds
Lead	0.315	0.126	Pounds

The costs of pollution and how much is gained by preventing it have not been sufficiently appreciated.[14] Nonetheless, logically, it should make sense that the more pollution we pour into our air, land, and water, the worse the problems we will eventually face and the more expensive the clean up. Conversely, the less pollution there is, the fewer problems we will have to deal with in terms of money and health.

Too Little Too Late

Among the writers who recognize an energy problem, there are those who believe alternatives could work, but express doubts that we have time to save ourselves. One of the most thoughtful articles is, "The Main Problem with Alternatives to Oil." The writer, who identifies himself only as Dave from Sidney Australia, makes some powerful points, among them:

> "Imagine all our civilization's energy is combined into a massive 'energy bank account.' ...Every ounce of fossil fuels that we then burn is a withdrawal from our account... The problem is that we have not earned this energy. We inherited it...Unless we invent another 'free' energy (which at this stage is science fiction)...we will run up a massive energy debt. ...Nature does not 'correct' an over-extended species kindly." [15]

In other parts of his article, he regrets that we didn't use oil as an ally in building an alternative energy infrastructure. So why not still use hydrocarbon energy for that purpose as I also advocate? He believes it is too late. His current pessimism is based on two premises: (1) that there will be a peak in oil supply soon with dwindling supply afterwards and 2) that the amount of energy we have left is already too little to build the alternative energy infrastructure we need.

What his pessimistic conclusions fail to take into account is again our huge coal reserves. However, the longer we delay regaining our energy independence the more likely it is that we will face massive economic disruptions consistent with the statement that "Nature does not 'correct' an over-extended species kindly."

The Wrong Alternative

Those who believe we still have enough time to correct our energy problems often worry that we are choosing the wrong alternatives to emphasize. Nuclear power is an example, so is ethanol with its poor

EPR, but hydrogen is perhaps the one alternative to hydrocarbons that is singled out most often for criticism as "the wrong alternative."

The most misplaced opposition to hydrogen is based on a powerful but misleading image. Some seventy years have past, yet the image of the Hindenburg still lingers. Just as concerns about home AC electrocutions and car gasoline explosions were overcome, so too can concerns about hydrogen be overcome. But imagined dangers can still persist, thanks to the compelling image of the Hindenburg crashing in flames.

Figure 31. The Hindenburg Disaster [16]

Ironically, recent analysis of the Hindenburg fire has revealed that the perceptions of the disaster are based on an incorrect assumption. Hydrogen leakage did not cause the crash. The skin of the Hindenburg had been treated with a substance that was both highly flammable and susceptible to accumulating static electric charge (a dangerous combination). The coating was added to make the skin stiffer and more durable. The fire was caused by a spark that ignited this skin. The Hindenburg would have burned and crashed in flames regardless of what gas it contained.[17] Hydrogen's combustibility, nonetheless, did not help. It accelerated the burn rate. Furthermore, deciding to create a blimp that depended on such massive amounts of hydrogen contained in a flimsy balloon-like structure was not very smart. It hardly compares with the manufacture of fuel cell cars or fuel cell stationary power generators that use much smaller amounts of hydrogen in well contained canisters.

A more powerful argument against hydrogen's use comes from those who believe a hydrogen economy is unrealistic. That is the position of Michael Behar in his paper "The Hydrogen Economy May be More Distant Than It Appears."[18] Behar thinks it is largely a waste of money, time, and precious resources to pursue hydrogen. He sees the emphasis on hydrogen as a major stumbling block to making real progress in defeating our hydrocarbon addiction. Behar provides a checklist of misconceptions and myths that he thinks have made a hydrogen economy more plausible than it really is. Among what he claims to be myths and misconceptions are:

1) We will never run out of hydrogen because it is abundant
2) You can drive hundreds of miles on a single tank of hydrogen.
3) If we mass produce hydrogen vehicles, they'd be affordable.

However, Behar's key argument for opposing these "misconceptions" is flawed. He claims that hydrogen's abundance is irrelevant because "We'll never get more energy out of hydrogen than we put into it." It's dangerous to say "never" when you're talking about technology. True, the energy it takes to separate hydrogen from oxygen using conventional electrolysis is more than the amount of energy produced. But through microbial fuel cells, researchers have already proven that hydrogen can be produced at a net energy gain. Furthermore, while electrolysis wastes non-renewable energy, which is a limited resource, it may not waste renewable energy. Producing hydrogen for backup generators may make sense if excess electricity is generated by wind turbines or solar panels. Storing excess electricity in batteries so electricity will be available when needed is already done, even though battery storage reduces the overall electrical output by 30%. Electricity can be generated directly to operate a household and to store a reserve of hydrogen because the overall EPR for renewable wind or solar generated electricity is very favorable and because only a portion of the electrical output has to go to storage.

Behars dismissal of the idea that we could get hundreds of miles on a tank of hydrogen is also flawed. Here he enlists JoAnn Milliken, chief engineer of the Department of Energy's Office of Hydrogen, Fuel Cells, and Infrastructure Technologies, to support his contention that "High-pressure hydrogen would take up four times the volume of gasoline...."

But Behar fails to mention metal hydride storage, hydrogen absorbing carbon nanotubes or sodium hydride balls, among other possibilities. These alternatives show promise of increasing hydrogen storage capacity.

Behar makes a stronger argument with his challenge to the idea "If we mass-produce hydrogen vehicles, they'd be affordable." It's not that mass production won't reduce prices. The problem is the availability of hydrogen. No one is going to buy a car that runs on a fuel they can't easily get. And no one is going to mass produce a car that people won't buy. Limited availability of hydrogen will limit car production.[19] So what would he do instead? He cites Physicist Amory Lovins, a hydrogen advocate. Dr. Lovins writes:

> "The most promising of these technologies is the gas-electric hybrid vehicle, which uses both an internal combustion engine and an electric motor, switching seamlessly between the two to optimize gas mileage and engine efficiency…Most experts agree that there is no silver bullet. Instead the key is developing a portfolio of energy-efficient technologies that can help liberate us from fossil fuels and ease global warming." [20]

The Valuable Insights of the Opposition

Besides a lack of consensus on what to do, we have to contend with a number of obstacles to the new technology that go beyond the technical challenges already cited. Like Behar's reasons for dismissing mass production of hydrogen cars, they are the most potent arguments against alternative energy brought up by skeptics. Without addressing them, we will have difficulty finding a way forward. Here without specific attribution is a list of the most formidable obstacles, followed by a description of each one. Every obstacle on the list has been acknowledged by the Department of Energy in one or more of its hydrogen plans. But most are generally applicable to alternative energy.

- Competitive disadvantage of alternative energy
- Lack of an alternative energy infrastructure
- Costs of transition
- Commercial immaturity of alternatives
- Lack of public support

Competitive Disadvantage of Alternative Energy. Because such a small percentage of energy comes from alternative energy sources, the cost of alternative energy is high. Because the cost of alternative energy is high, the prospects of increasing its market share appear low.

Lack of an Alternative Energy Infrastructure. Oil not only has a cost advantage, it has an infrastructure advantage. It relies on an existing, widespread, well-established infrastructure that was made for it. Alternatives don't have an obvious existing infrastructure, critics argue, and building one on the scale needed is going to be very expensive.

Costs of Transition. The cost of replacing our current energy infrastructure with an entirely new alternative infrastructure, as the government proposes, could be staggering. To succeed in transitioning to this new energy infrastructure, we potentially have to bear the cost not only of creating it, but of maintaining the current one until we are well into the transition. Unfortunately, even maintaining the current infrastructure is going to be costly because we've put off so many problems. The installed capacity of our power grid, for example, is not increasing at a pace that sufficiently addresses increased demand. Here is what the Department of Energy has to say about that problem:

> "By the time wholesale electricity competition began in the United States in 1996, utility investment in power plants had slowed considerably...The resulting lack of generation growth has led to tight electricity supplies in much of the United States."[21]

We are not only failing to build power plants fast enough, we are failing to increase the number of transmission lines quickly enough. Consequently, we are becoming more vulnerable to blackouts. Why? Transmission lines have a limit to how much current they can transmit. As transmission lines carry more and more current, they get closer to that limit. Our grid system is supposed to be failsafe. It is redundant. And even through it is monolithic, it has built-in modularity to isolate problem regions. There are usually two or more transmission paths between any point A and any point B, which makes it possible to shed current when there is too much or to shunt current from lines carrying too much (or which have failed) to lines carrying less. But what happens

WHAT HOLDS US BACK   219

if the transmission lines receiving additional current are constantly overloaded?

We have already experienced a major failure on the eastern portion of our monolithic grid system in 2003 when a key power plant shut down. The chronology of events was due in that case to some poor decisions that allowed too much power to reach the lines. But it suggests the dangers we could face in the future if we don't address the problem of sufficient growth.

August 14, 2003
1:58 p.m. Eastlake, Ohio. The power plant owned by First Energy suddenly shuts down. It had been experiencing extensive recent maintenance problems. Meanwhile demand is increasing. It is summer and demand tends to peak around 3:00 p.m.

3:06 p.m. South of Cleveland, a First Energy transmission line fails.

3:17 p.m. The Ohio portion of the grid experiences a voltage drop. The dip is temporary, thanks to a shift of power from the failed First Energy transmission line to another line. Controllers don't react.

3:32 p.m. The line which received First Energy's transmission line power overheats, sags into a tree, and goes off line. The Independent System Operator (ISO) who supervises the lines in the Midwest calls the First Energy controllers in an effort to determine why the lines failed but doesn't alert system controllers in neighboring states.

3:41 and 3:46 p.m. Two breakers that connect First Energy's portion of the grid with that of American Electric Power are tripped. Breakers are designed to trip when they receive too much power. By tripping, they isolate and protect the grid from local problems.

4:05 p.m. Tripping and shunting causes a sustained power surge on some Ohio lines, signaling that problems are growing.

4:09 p.m. Ohio draws 2 gigawatts of power from Michigan.

4:10 p.m. More breakers begin to trip out, first in Michigan, then in Ohio, blocking the Eastward flow of power. In seconds power surges increase, automatically tripping breakers and taking East Coast generators off-line to protect them. Generators go off line, and the blackout is on in much of the Northeast U.S and parts of Canada.

Justifying the additional expense of creating a new infrastructure might not be easy in the presence of our looming problems with the current infrastructure.

Commercial Immaturity. Advances in alternative energy technologies are uneven. The relatively greater maturity of hydrogen fuel cell cars compared to hydrogen production techniques is an example. Retarding progress further, the primary emphasis is not always on developing the most promising solutions. Politics affects decisions.

Lack of Public Support. To date, there is not nearly enough public support for alternatives. Fears of global warming have galvanized some to act but hasn't yet carried the entire country. Nor have the issues with imported oil. Why? Our country's economy today is relatively stable, but there is a looming recession. Local and personal problems, including an increase of personal debt, may seem more pressing and can lead to an "everyone-for-himself" mentality, a sense that there isn't much we can do individually to make a difference about the wider problem of energy.

Given this mentality, people are likely to argue, "Maybe the air is a bit unhealthy to breathe; maybe dependency on imported oil isn't such a good idea given what's happening in the world; maybe gasoline prices are getting a bit high, but what can I do? How do I get to work in the morning without gasoline? How can I heat my home without oil? So maybe there are alternatives, but why should I pay more for alternatives that probably aren't that good anyway?" In the absence of a huge energy crisis, getting people to change their energy consumption habits and to work together for energy independence and the reduction of pollution will be a challenge.

Other Factors Holding Us Back

To the obstacles already mentioned, I would add some other social and political factors that affect perceptions, retard progress, and depress the motivation to change.

WHAT HOLDS US BACK

- Price fluctuations
- Subsidies
- Pressure to keep oil prices low
- Oil companies' mixed signals
- Failure to involve the market
- Unrealistic energy budget and the war
- Too little state involvement in the past

Price Fluctuations

Over the last 30 years, if Americans had paid the much higher price for gasoline and electricity that Europeans have paid, it is hard to imagine that we would have made so little progress towards a solution. When oil prices have gone up, for example, the public has responded by seeking ways to save money, which has translated into more efficiency and less waste. When oil prices are low, we tend to be wasteful and inefficient, no matter what the long-term consequences may be.

This behavior is obvious to anyone who examines what has happened as oil prices have fluctuated over the last 30 years. After the oil embargo of 1973, the price of oil climbed, as indicated in Figure 32.

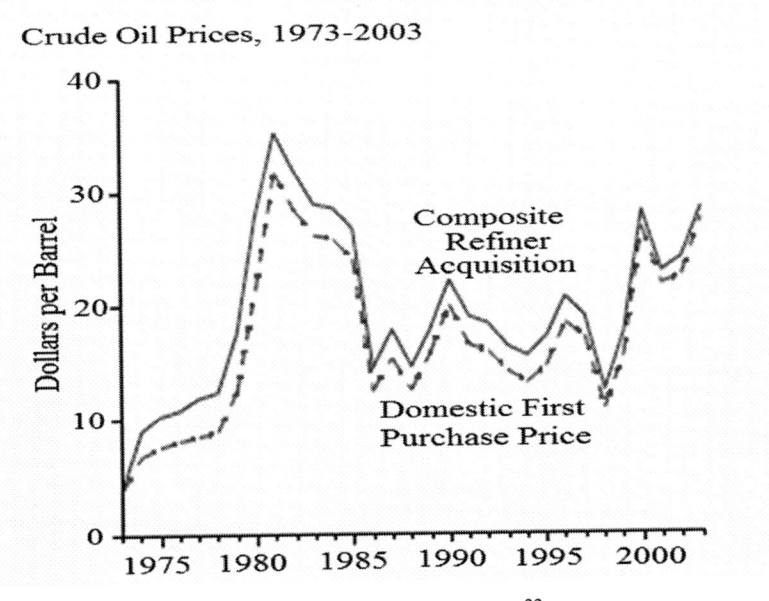

Figure 32. Crude Oil Prices[22]

The average **domestic first purchase price** is the average price at which all domestic crude oil is purchased.[23] The average cost of crude oil to U.S. refineries (referred to as the **composite refiner acquisition cost**) greatly effects the final cost of petroleum products.[24] These two indicators track the price of crude oil.

Thanks to the gasoline price increases, the United States under Jimmy Carter's presidency (1977-1981) made considerable strides in improving the energy efficiency of cars and electrical appliances. We still have the legacy of his efforts today. However, from the mid 1980s through the late 1990s, as Figure 32 indicates, OPEC dramatically dropped its price for oil. The alternative energy industry, which had gotten a foothold in the energy market in the late 1970s, shrunk dramatically. In contrast, the number of SUVs rose dramatically.

Who benefited from the increase in SUV sales? The oil companies benefited because SUVs need more gas than regular cars. The car companies benefited because bigger vehicles generally carry bigger sticker prices and yield bigger profits. And consumers, influenced perhaps by a powerful marketing campaign, clearly felt they benefited since they bought SUVs in record numbers. In a market driven economy, this situation makes failure almost inevitable for manufacturers of alternatives.

Subsidies

So why doesn't the government just stop subsidizing oil exploration by oil companies, which are glutted with profits from rising oil prices? Why doesn't our government let prices rise so alternatives can compete? One explanation is that the more oil that the oil companies find and exploit within and along the coasts of the United States, the less dependent we are on Middle Eastern countries and other unstable, unfriendly foreign countries.

The converse is also true. If oil companies shut down their wells in the U.S., which some have threatened to do if subsidies are withdrawn, we become more dependent on foreign oil more quickly.

In other words, the government is caught up in another aspect of the transition problem. It feels it has to maintain domestic oil production as part of the current infrastructure to avoid a devastating disruption of the economy while it tries to create an alternative infrastructure – all of this while conducting a war against terrorists who could disrupt oil supplies.

Pressure to Keep Oil Prices Low

Naturally, government interventions that raise oil prices are resisted by the public, which suffers from paying more. So we try to hang on to the status quo while our position continues to erode. At the same time, we reach for explanations for what is happening based on half truths. We see high prices for gasoline and hear about record oil company profits. What conclusion are we likely to reach? The oil companies are gouging the public.

That conclusion ignores the root causes. Increases in the price of oil are a symptom of tightening supplies and instability in oil producing countries. If not addressed, resulting supply constrictions will make us more and more vulnerable to oil commodity monopoly blackmail and worse.

So what are we to do? Gasoline and home heating oil price rises over which we have little control should not be the prime motivator of action, but they clearly are. If we keep gasoline prices as low as possible, we forestall action, because without high gasoline prices, most of us aren't compelled to take action to reduce our gasoline consumption. But if prices are raised, there will be economic hardship. We are caught in a dilemma, with no obviously good options. It is hardly surprising we are having such a hard time making progress. This unfortunate dynamic is a consequence of our dependency on only a single vital commodity, oil. Wait until we face a natural gas shortage as well.

Oil Companies' Mixed Signals

Making the current situation even more complicated are the divided opinions within the oil industry. British Petroleum is one major oil company that has expanded its business model to include solar and wind energy, has contracted with major universities to do research in alternative energy, and seems to be genuinely committed to going "beyond petroleum." Exxon Mobile, in contrast, sees little future for alternatives, judging by its statements. In a report that was delivered outside the U.S. (in England), Exxon Mobile stated:

> "We expect oil and gas will continue to be the primary sources of energy through the outlook period, at a stable share of about 60 per cent. Oil demand growth is projected at 1.5 per cent per annum. Natural

gas growth is about 2.2 per cent and reaches a 25 per cent share of total energy by 2030.   Finally, we expect wind and solar growth of about 10 per cent driven by subsidies and related mandates. Even so, by 2030, their share will only begin to approach 1 per cent of total energy." [25]

If Exxon Mobile doesn't expect much of solar and wind energy, it is even more pessimistic about the prospects for hydrogen: "...hydrogen vehicles do not emerge as a significant factor even in our 2030 outlook." It is hard to interpret this as a declaration of faith in the government's hydrogen vision. Exxon Mobile's projection that alternative energy of any kind will have a market share no more than 1% by 2030 is hardly in line with the objectives or projections of the government's *Hydrogen Posture Plan,* much less with the government's vision of a hydrogen economy or any alternative energy economy.

With such diverse attitudes towards alternative energy, the oil industry is sending contradictory signals.

A Failure to Involve the Market

Our government's primary emphasis, which has been to build a hydrogen economy, runs into difficulties not just because of the hydrogen storage and production problems, but because it fails to effectively use the market, despite government perceptions to the contrary. The government plan harbors an underlying belief that it must create an alternative infrastructure based on hydrogen first. The government strategy is telling, "Create a new infrastructure and they will come." Insufficient thinking has gone into creating a real market basis for this infrastructure.

Their plan hasn't anchored its strategy in the market, but in the government as the first user. And even though the Department of Energy has sunk billions of dollars into research on alternative energy from solar panels to ocean wave generators, the *Hydrogen Posture Plan* emphasizes the use of solar and wind power mainly for one purpose, the creation of hydrogen.

This plan for our energy future is expensive in its approach and too narrow in its vision. It leaves everything other than hydrogen largely outside of this plan for energy independence. Excluded from the plan, for example, is the use of solar and wind energy to generate electricity for the stationary power sector and the use of monetary incentives to encourage the public to buy hybrid cars and energy efficient appliances.

Furthermore, outside the hydrogen plans, there seems to be no overarching systems vision that could optimize what we are doing with alternatives. Instead we have a "machine gun approach" that counts on scoring a hit by shooting so many little monetary bullets that at least one is bound to hit something. This is a good approach for introducing innovation but not for improving existing products. Competition is a better stimulus for making improvements. Nor is there an adequate contingency plan if we experience significant oil shortages for more than a few months.

An Unrealistic Energy Budget and the War

The 2007 Energy Bill committed money to a "solar initiative," but preferred to commit more money to continue building a new hydrogen energy infrastructure virtually from scratch and to producing corn ethanol despite its low EPR and meager prospects of fulfilling by itself our transportation needs. Then there is the diversion of funds to the Iraq War. By September 2007, according to government figures, we had spent something like 500 billion dollars in Iraq. Democrats claim it is more like one trillion dollars!

Too Little State Involvement in the Past

It would be reassuring to discover that the states have been thinking about energy for the last 30 years and have secretly come up with a more compelling plan. After all, self-reliance is ultimately a local matter for the state, the community, and individuals to work out. Unfortunately, for decades only a few states thought much about the problems of hydrocarbons or about what they could do to reduce our dependency on oil imports. Energy policy was perceived to be the responsibility of the federal government.

So there are lots of reasons why we have made so little visible progress. But there are also reasons why change is inevitable. Today, oil has reached prices that can not be ignored; turmoil in the Middle East has increased to an extent that the dangers are obvious, and the Intergovernmental Panel on Climate Change (IPCC) has been able to produce increasingly strong evidence of our involvement in global warming. With these developments, many states are making energy a priority. They have begun to act together in groups of states and individually. With their greater participation, with an awakening press, and with dedicated individuals, advocacy groups, and world leaders demanding action, the momentum necessary for change to occur is noticeably building.

PART THREE

The Age of Transition

"We in America do not have government by the majority. We have government by the majority who participate."

Thomas Jefferson

"No matter what we attempt to do, no matter to what fields we turn our efforts, we are dependent on power. We have to evolve means of obtaining energy from stores which are forever inexhaustible, to perfect methods which do not imply consumption and waste of any material whatever. If we use fuel to get our power, we are living on our capital and exhausting it rapidly. This method is barbarous and wantonly wasteful and will have to be stopped in the interest of coming generations."

Nikola Tesla

24

California Bellwether

One of the first states to take the initiative with a policy and plan of its own was California. Its impetus stems in part from its liberal base (which includes some of the most environmentally conscious groups in the country), from its problems with air pollution, from its growing problems with refineries and oil import costs, and from its experiences with Enron and the rest of the crowd that almost bankrupted the state's three principal energy utility companies. The California experience is possibly the clearest warning yet of the kind of threats we face and the even greater dangers of an ineffectual response. It is worth examining what happened and what has happened since then, as California went from being ineffectual to being an effective energy trend setter.

A number of factors converged to place California in an energy crisis like nothing most of us had ever experienced before. The problems seemed to begin in the summer of 2000, but they actually originated at least four years earlier when Governor Pete Wilson partially deregulated the price of electricity.

What his partial deregulation meant was that the price of electricity on the spot energy market would be allowed to float while the price the utility companies passed on to their customers would be capped. The decision to move to partial deregulation had the unanimous support of the California legislature. Apparently, no one considered the possibility that the floating spot market price might rise above the customer price cap.

As part of the partial deregulation strategy, the three principal utility companies in the state were forced to divest themselves of 40% of their installed capacity, but they still retained responsibility for electricity distribution.[1] The divested capacity was sold to "independent power producers" (energy traders) which included Enron. Already concerned about clean energy, California had also contracted for hydroelectric power from Oregon and Washington for much of the out-of-state electricity it might need during peak power usage.

Unknown to California's energy regulators and lawmakers, these factors had provided the perfect situation for unscrupulous energy traders to prey upon the state with what the traders believed was impunity. Their chance to pounce came in 2000, thanks to summer heat waves coupled with the effects of a prolonged drought. Energy reserves within California that summer began to get low during peak hours. Oregon and Washington hydroelectric power providers, also affected by drought, were hard pressed to alleviate the apparent shortages. Although there is no evidence that demand actually exceeded supply, California began experiencing rolling blackouts. Among the most severely hit was the San Francisco Bay area on June 14, 2000 when 97,000 people lost power during a heat wave.

Fearful of further shortages, California's utility companies began to buy energy from wherever they could on the spot market, at prices that became increasingly more exorbitant as Californians grew more desperate. Lowering demand by just 5% would have resulted in a 50% price reduction during the peak hours of California's electricity crisis of 2000 and 2001, according to one study,[2] but plans for a coordinated response to the crisis didn't exist. The state was caught completely unprepared.

Unable to pass on the rising costs to their customers because of the price cap, the utility companies warned they were heading for bankruptcy. Alarmed, on January 17, 2001, our new Governor Gray Davis declared a state of emergency and set about spending millions of dollars in support of our beleaguered utility companies. It did little good. By April 2001, blackouts were affecting over 1.5 million customers, and despite attempts by California to prop it up, Pacific Gas & Electric began bankruptcy proceedings.

With California facing another bleak summer of increasingly frequent blackouts early the coming year, Governor Davis signed long term contracts for energy at prices that could hardly have been envisioned just a year before. Lacking confidence in California's ability to regulate electricity, many of California's larger businesses, including Compaq Computers and Prudential Insurance of America, signed separate energy contracts of their own at exorbitant prices.

In effect, the state's attempt to foster competition in the energy market had inadvertently left everyone in the state vulnerable.[3] I want to re-emphasize that while California's energy infrastructure was stretched, that was not the immediate cause of the crisis. The immediate cause of the crisis was a contrived shortage that was made possible because of California's stretched energy infrastructure, the spot market, and the ability of unscrupulous energy traders to gain partial control of the power grid as the result of our utility companies' divestitures.

The perpetrators of this crisis believed their machinations would remain buried beneath the chaos they created and the intricacies of energy trading on the spot market, despite attempts by the utility companies to expose them. The confidence of the traders was not unfounded. If Enron had not been driven into bankruptcy by its corrupt practices, it is questionable whether the federal government would have obtained conclusive proof of wrongdoing. It is not even certain that the effort to prosecute Enron would have been made. With Enron's bankruptcy, however, came the disclosure of the now infamous Enron tapes and concrete evidence of what really happened.

As the tapes revealed, Enron's traders inventively "gamed" the energy system under code names such as "Death Star," "Ricochet," "Ping Pong," "Fat Boy," and "Black Widow". A key strategy was to create phantom congestion using their partial control of the power grid. To create phantom congestion, Enron first obtained transmission rights on a key transmission line, a choke point, connecting Northern California to Central California. Then Enron traders began to clog this transmission line with power, essentially diverting power from where it was needed. As one trader is heard saying on tape,

> "What we did was overbook the line we had the rights on during a shortage or in a heat wave. We did this in June 2000 when the Bay Area was going through a heat wave and the ISO [the Independent System Operator who supervises the lines] couldn't send power to the North. The ISO had to pay Enron to free up the line..."[4]

The price for freeing up the line was several tens of millions of dollars.

What Enron did was quickly copied by others with a gleeful maliciousness that had one trader joking about their control. The following are excerpts from the Enron tapes, which capture the attitude of the traders. "This is where California breaks. Yeah, it sure does, man...You guys need to pull your megawatts out of California on a daily basis. They're on the ropes today. I exported like a f-----g 400 megs.

Wow, f__k 'em right! You want to do some fat boys or, or, whatever, man, you know, take advantage of it."[5] "Fat boy" was another scheme Enron traders used to drive up prices.

Gray Davis's attempts to shore up the utility companies were fully anticipated by Enron's executives who were wantonly destroying the utility companies' credit. Skilling, one of Enron's executives, described their strategy this way, "As the [utilities'] credit exposure gets too high, we will limit the amount of power we deliver into California. Eventually, the state is going to have to provide these companies with the credit support from somewhere to support their purchases."[6]

An Enron Energy Services (EES) executive Steve Barth allegedly engineered another part of the strategy, which was designed to force California's big commercial companies into signing those separate energy contracts. It was later described in these terms: "First our traders are able to buy power for $250 in California and sell it to Arizona for $1,200 and then resell it to California for five times that. The EES was able to go to these large companies and say sign a 10 year contract with us and we'll save you millions."[7]

I could continue, but the picture should already be clear. One reporter looking back on what happened remarked that with American companies like Enron, who needs terrorists[8] or foreign cartels?

California's energy problems brought about a rare recall election which pushed Gray Davis out of office and replaced him in October, 2003 with what must have seemed to the rest of the country like a very strange choice, a movie star, who also happened to be a Republican in a heavily democratic state, Arnold Schwarzenegger.

Schwarzenegger had never run for any office before, but after he became governor he showed two indispensable qualities that California needed. He is a problem solver and a unifier. Under his leadership, California began to address its energy problems as never before. To start with, California's government acknowledged that the problems we faced were not solely due to market manipulation by Enron and friends. California's government recognized that the energy crisis had root causes as well as immediate causes.

A major root cause was limited supplies in the presence of a growing population. The government also recognized that the problems extended into the transportation sector, which is even more vulnerable.

As Terry Tamminen, then the secretary of the California Environmental Protection Agency (EPA), said in a 2003 interview:

"[We have] 36 million people and our population is growing by almost 600,000 every year. We have 30 million motorized vehicles in the state, almost one per person. And the vehicles in showrooms today have worse fuel economy than in 1987."[9]

So what is California's answer to the growing problems with energy supply in both the stationary power and transportation sectors? Schwarzenegger's initial response was to continue building the hydrogen highway, which he envisioned as the replacement for the hydrocarbon highway. Like federal plans, this is the "create a new infrastructure and they will come" strategy. Despite the possibility of inadequate supplies of hydrogen, the California plan emphasized building fueling stations and obtaining fuel cell cars that because of their relative maturity make a near term hydrogen economy seem more plausible. In support of this strategy, acts were proposed to the California legislature. Examples are:

The Hydrogen Act referred to House Committee on Science – Subcommittee on Energy, March 25, 2003. In it, the state of California proposes to require automakers to develop a plan by 2015 to make fuel cell vehicles (FCVs) available to a mass market.

Fuel Cell Vehicles AJR50. It affirms California's commitment to using hydrogen and fuel cell technologies to achieve a clean transportation economy.

H2 GROW Act HR1180/5. This bill provides incentives for purchasers of hydrogen fuel cell vehicles and encourages hydrogen fuel sales through credits and tax breaks.[10]

Because this California Plan is much more specific than the federal plan, it is easier to assess. Here are some of the potential problems:

- A series of service stations are being put in place with no certainty that production, delivery, or storage demands can be met. Limited supplies of hydrogen will limit the number of customers and inhibit growth.
- By creating a spread out network called the hydrogen highway, the California plan is in danger of spreading itself too thin. Distributing

stations across a network will exacerbate the problem of hydrogen supply and demand.
- It has no assurance that the public will participate since there is little initial customer base other than the state government.
- Hydrogen is dispensed by very energy inefficient methods (as a liquid).
- Cost estimates are based on too little data.
- The hydrogen highway initiative creates a separate energy infrastructure parallel to the existing one, which is very expensive.

California's hydrogen highway plans had many of the same flaws found in the federal government plan. It did not leverage the existing infrastructure effectively to promote transition. It did not expect a hydrogen economy to grow naturally, driven by market profit incentives as the hydrocarbon energy infrastructure had. It did not include adequate contingency plans in the case of a sudden oil shortage.[11] In short, the hydrogen strategy developed in California, like that of the federal government, was not encouraging.

Fortunately, Governor Schwarzenegger did not stop with the hydrogen highway initiative. Over several years, he and his supporters have moved forward with other initiatives. We now have biofuel, solar, and wind initiatives, better recycling strategies, along with legislative bills to improve hydrogen production and the hydrogen highway concept, and a host of other initiatives designed to make California the foremost green state in the nation. As a consequence, California is gradually expanding its energy base. At the same time, California is increasingly integrating its energy plans within a master plan that more recently added a contingency plan to deal with potential serious energy shortages.

More than anything else, California's governor has galvanized the environmental groups, attracted a number of environmental activists to his side, and won the overwhelming support of the Democratic legislature for his efforts, which has unified California's official response. California's commitment to alternative energy has clearly influenced other states and may well influence the federal government.

25

The Future in Our Hands

There are powerful reasons for establishing an alternative energy economy. If we succeed in making the transition, we can expect to have:

- National energy self-sufficiency
- An increase in national security
- Significantly less international debt
- A significant increase in American manufacturing jobs[1]
- A new energy industry[2]
- An improved environment
- Health improvements as man-made environmental toxins are reduced
- A reduced threat of man-induced global warming

While it can not be overstressed how important it is to have a leader like Schwarzenegger who is both a problem solver and a unifier, neither his plans nor those of the federal government seem to me to be sufficient. Because of their energy infrastructure replacement strategy and because these plans are too narrow in their emphasis on hydrogen and still too piecemeal otherwise, we can easily fritter away enormous amounts of money and time if we follow them.

Is There a Better Mousetrap?

The plan I propose takes into consideration the current status of our technology and the obstacles to alternative energy I've cited. It includes much of what California and the federal government are now doing but uses a more focused systems-engineering approach to improve their plans. In presenting my approach, I am going to start with a simple **guiding concept** – integrate the stationary power and transportation sectors using electricity.

Use a Guiding Concept

The key to understanding how to proceed and what to emphasize can be seen in Figure 33.

Energy Breakdown

Stationary Power Sector: Oil 1.6%, Other 3.1%, Hydroelectric 7%, Natural Gas 20%, Coal 49%, Nuclear 19.4%

Transportation Sector: Gasoline 98%

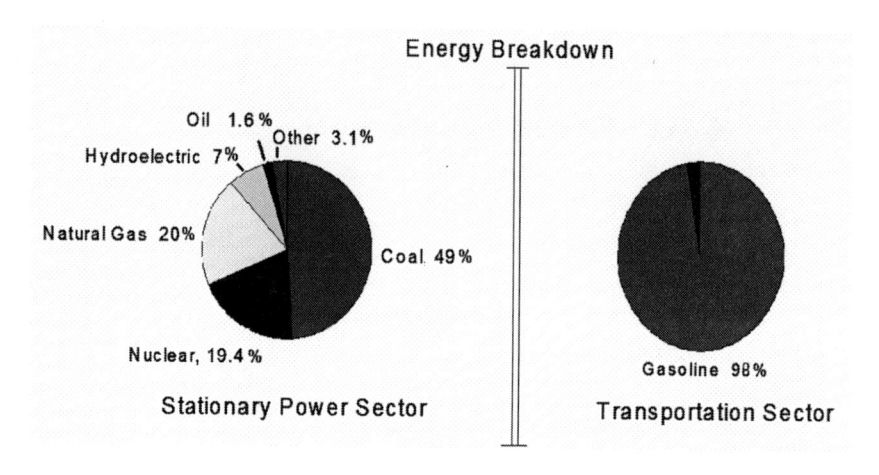

Figure 33. Sources of Energy for Stationary Power and Transportation[3]

From Figure 33 it should be clear that we have a lot more energy options in the stationary power sector than we do in the transportation sector. The availability of natural gas, hydroelectric power, nuclear power, a rising percentage of renewables such as solar and wind resources[4] and our most abundant hydrocarbon, coal, gives us a variety of choices for generating electric power. As a result, less than 2% of the energy we use in the stationary power sector comes from oil.

In contrast, the predominance of the gasoline internal combustion engine limits the choices in the transportation sector much more severely. Its predominance has ensured that gasoline is used in around 98% of the cars on the road. The transportation sector is also the sector that produces the most pollution. Given these facts, it may seem logical to focus on solutions in the transportation sector. After all, reducing oil imports by using other sources of energy can both reduce pollution and free ourselves from the energy vise we are in.

The problem with this focus is that the alternative energy options for transportation are too limited and weak at present. The most obvious strategy would be to find alternative fuels such as hydrogen and ethanol. But in fact, the key to succeeding is to connect the stationary power sector to the transportation sector by using stationary power to recharge electric and hybrid vehicles. Through an electric plug, there would effectively become available to the transportation sector, coal, nuclear energy, hydroelectric power, solar energy, geothermal energy, wind energy, wave energy, and alternative fuels. In return, vehicles could one day provide power back to the grid. With integration of stationary power and transportation, we reduce our dependency on foreign oil. And by gradually converting the stationary power sector to clean alternatives using clean conversion devices such as wind turbines, solar cells, and fuel cells, we make both the stationary power and transportation sectors increasingly pollution free. This simple guiding concept – integrate the transportation and stationary power sectors using electricity – can give our efforts focus, discipline, and empowerment.

Use a Systems Engineering Approach

One reason this simple guiding concept is so powerful is that it encourages a systems engineering approach to solving energy problems. A systems engineering approach is one that starts with the "big picture." It determines how a system as a whole should operate through the interaction of its major components. Beginning at a high level, it then works out how the system components should interact in greater and greater detail.

This approach is comparable to getting a high level view of terrain you want to traverse before you traverse it. The guiding concept presents a view of the power grid and the transportation infrastructure as a single system. A systems engineer can therefore ask, how must the components of this system interact to make the system as a whole work? This is a very different approach than focusing on improvement of individual components. Focusing on individual components fosters a tendency to neglect the overall system, and can put too much emphasis on individual components as "the answer," which reduces the chances of finding a workable solution.

Harness the Market to Accelerate Improvements

To support rapid and efficient integration of stationary power and transportation, we have to **harness the power of the market**. Products for which the government is the first user but the public is the eventual intended customer haven't had a history of improving very rapidly. On the other hand, products exposed to market competition improve quickly or they are set aside.

Among many reasons for this difference between the government as the only customer and the open market is practicality. When products are used in an open market, their flaws and their potential for improvement become more apparent. But the market is driven by profit, and alternative energy devices and fuels are not yet profitable enough to win a large number of customers. That is why the government has enlisted itself as a first user.

But is that the only option? I don't think so. Many companies already use another option. They enlist customers who are willing to use early versions of products and to provide feedback that helps rapidly improve the products.

If we can improve alternatives and give them a firmer foothold in the market by driving down costs while helping them gain profitability, we can bring virtually everyone in the country on board, even those in opposition who may now see their interests threatened. If there is a profit to be made, if prices are competitive, if momentum grows, the competitive disadvantage of alternative energy will evaporate. We won't lack eager investors or buyers who will overcome any opposition. If we fail to make alternative energy commercially viable, and resort to the usual recourse of self-flagellation and blame we will leave most people dispirited and less motivated to make changes.

Evolve Rather than Replace the Existing Infrastructure

Back in 1906, automobile manufacturers were able to take advantage of an existing oil infrastructure, initially geared to produce kerosene. By using that infrastructure to produce gasoline instead, they were utilizing the existing infrastructure to create something new. By using an existing infrastructure to promote change, they reduced the costs of transition to the new infrastructure.

We can do something similar, because an infrastructure that alternatives can use already exists. We can generate electricity from multiple, renewable sources that can be plugged into the existing grid. And we can use bridging technology that fits within our current infrastructure. Both actions will help us to evolve the existing infrastructure into one for alternative energy. Evolving an existing infrastructure into a new one is more sensible and economical than trying to entirely replace the existing infrastructure using the government's "create a new infrastructure and they will come" parallel development approach. But it will be challenging.

Establish Plan Components

To the guiding concept of sector integration and systems approach, I add a number of plan components that can support them.

Plug-in Hybrid and Electric Cars

First I add the plug-in hybrid car. As they become commercially viable, I would also add all-electric cars (including fuel cell cars). They are vital components of the plan because they make it possible to connect the transportation and stationary power sectors through electricity

Clean Energy Electric Generators

To the stationary power sector I add fuel cell - gas hybrid generators and electric generators that use solar, wind, wave, and geothermal energy. These generators will help make both the stationary power and transportation sectors increasingly pollution free. They also make possible a distributive power grid based on renewable energy.

Alternative Fuels

Next I include the judicious use of renewable biofuels and hydrogen. For a few states, it may make sense to use biofuels and hydrogen from the start as transportation fuels. But for the majority, I think biofuels and hydrogen should first be brought into the stationary power sector where their use as a limited resource is likely to be more efficient and cost effective, and where they are likely to be more profitable. For example, based on what is already possible, using a hybrid consisting of a fuel cell and a gas turbine to generate grid electricity would be more cost effective and profitable than producing fuel cell cars, at this point. The same could be said for biofuel use for stationary power. If we can increase supplies and improve EPR while keeping the production of biofuels and hydrogen environmentally friendly, we can expand their use into the transportation sector.[5] Expansion for biofuels would mean making plug-in hybrid electric cars into flexible fuel cars.

With plug-in hybrid flexible fuel cars, we attack the problem of oil in the transportation sector from both the stationary power sector and the transportation sector so long as the EPR for the biofuel is favorable.[6] We also tackle and resolve the catch-22 situation posed by critics who have argued that without cars that can run on alternative fuels, alternative energy fuel stations have no economic incentive to operate, and without alternative fuel stations, cars that can run on alternative fuels have no place to go. Plug-in hybrid flexible fuel cars are a solution to this catch-22 situation. They are able to run on gasoline when biofuel and electrical outlets are not available. We can create hybrid cars that get 100 miles per gallon of gasoline because their energy is supplemented with plug-in electricity stored in batteries and then further reduce the amount of gasoline consumed by blending the gasoline with biofuels. These cars can fit within the existing energy infrastructure and help it evolve into a new integrated infrastructure.

Alternative Energy Zones

To speed up the development of a new alternative energy infrastructure and rapidly mature alternative products, I add the creation of alternative energy zones to the components list. An alternative energy zone is a small geographically defined area in which a community of customers uses the "guiding concept" to create alternative energy for their stationary power and transportation needs. Alternative energy zones are a means of building profitable alternative energy companies through a solid customer base.

Modeling and Measuring for Success

To ensure progress in the alternative energy zones, I add the use of systems engineering models, simulations, baselines, measures and metrics, to the components list.[7] What I mean by a systems engineering model is a paper or miniaturized construct of a system. A model is similar to a simulation in that both suggest how a system should function. But a model is static. A simulation is dynamic. Today simulations can mimic a system's operations using the "virtual reality" capabilities of a computer. Both models and simulations should be used to guide the construction of an actual alternative energy zone.

Once established, the alternative energy zone can then be monitored to see how it actually functions. Its initial state can be captured to establish a baseline by measurements and metrics. Measures are values associated with single attributes of a system or product. Metrics are comparative values. Baselines are snapshots of a system's state at a particular point in time.

As it grows and develops, the zone and its products can be measured against a baseline to determine improvements as you might measure a company's product improvements in quality over time. In an alternative energy zone, it is easier to create models and simulations and to apply them than it is in the wider economy because a zone is smaller and less complex yet representative of the wider user community.

Some sample high-level measures that might be applied are these that characterize an alternative car monitored in an alternative energy zone:

Distance on a Tank of Fuel. The distance an alternative car can go on a "tank of fuel" (hydrogen, biofuel, electricity from batteries, etc.).
Goal: travel 300-350 miles.

Safety Record. Automobile accident fatalities.
Goal: Fatalities equal to or preferably below that of conventional automobiles.

Price of Energy. Price of fuel or electricity to run a car.
Goal: The price should be below that for conventional cars in their class.

Durability. Time until car parts wear out.
Goal: Durability should be equivalent or better than that for conventional cars in their class.

Reliability. Ability to consistently function over time as advertised.
Goal: Alternative cars should be at least as reliable as conventional cars.

Convenience. Closeness of fuel supplies to the customer; distance to fueling stations, etc. Goal: Access to alternative "fuel" should be as convenient as it is for conventional cars. If fueling stations are used, for instance, they should be within five to ten miles of work, home, market, and school. The effect of proximity can be measured in relationship to customer satisfaction.

Some sample metrics are:

Ratio of Output Energy to Input Energy – EPR. Example: determine how much energy an energy conversion device (such as a solar cell) generates over time versus how much energy it takes to manufacture or produce it.

Pollution Caused by Alternative Energy Versus the Pollution Caused by Comparable Hydrocarbons. Comparison of pollution levels emitted by alternative energy conversion devices and levels emitted by comparable devices which have a similar function but use hydrocarbons.

Modeling, simulation, baselines, measures and metrics can help determine what works and what doesn't; what causes the most and least pollution; and what is most cost effective and least cost effective within an alternative energy zone. The predictive power of measures and metrics, models and simulations can guide companies, universities, national labs, and alternative energy zone directors in planning and fostering rapid improvements that optimize alternative energy products for a wider market. Alternative energy zones will be the focus of Chapter 27.

Innovative Thinking Techniques

To complement the establishment of alternative energy zones, I add approaches to problems that promote innovative thinking. We need to seek out new solutions to new problems. The tendency today, on the contrary, is to inappropriately apply old solutions to new problems. If we are going to reach the goal of an alternative energy based economy, we need a lot of creative and innovative thinking. It is not a capability we have ever noticeably lacked, but it needs to be better directed.

One technique for stimulating creativity is to imagine that we have reached a goal, then trace backward from the goal to see what could have produced the desired result. Another method is to look into very different fields for new paradigms. As James Burke has pointed out in his book

Connections, solutions to problems in a particular field often come from fields that don't seem to be related. We need to be able to think "outside the field."

To encourage progress, we need energy company involvement. Increasing their stake in alternative energy reduces their opposition to it. And given their energy expertise and creativity, the traditional energy companies are a natural choice to produce oil from coal, as well as to find ways to efficiently produce hydrogen, solar and wind energy conversion devices, and alternative energy fuels such as ethanol and syngas in the quantities that will be needed. Some oil companies, in fact, already have alternative energy divisions. Most are already aware of the problems with oil and realize they should transition to alternatives someday. But some like Exxon Mobile are disdainful. If it begins to make economic sense to create alternative energy, however, that attitude will change or they will be left behind by competitors.

Systems Grants and Subsidies

To further alternative energy zones, I add to my components list systems research grants and customer subsidies. Alternative energy zones embody a systems engineering approach in that they are models of an entire working system. System grants are grants that would aim at improving the working alternative energy system. The small-scale alternative energy economic zones should quickly pinpoint strengths and weaknesses, and determine the scope and impact of problems. System research grants can be targeted to enhance strengths and address uncovered problems, taking into practical account the potential effect of improvements on the entire system and the most promising solutions. Customer subsidies within alternative energy zones can entice customers to participate.

Energy Efficiency and Clean Energy with Computer Control

Another component of my plan is the promotion of energy efficiency as Carter did in the late 1970s. As we increase dependence on our stationary power sector in order to reduce dependence on petroleum in the transportation sector, we need to compensate for the shift. Remember that shifting the energy burden from the transportation to the stationary power sector reduces expenses. The cost of energy in the stationary power sector is well below that of the transportation sector and should remain so, thanks to its ability to use our abundant coal supplies and to implement increasingly cleaner more varied energy options. But we have to do more.

Increasing the energy efficiency of home appliances, using solar thermal energy, which can be used for both passive solar heating and power generation, using better insulation, and using computers to regulate and improve home energy efficiency are crucial elements of this plan. If homeowners need to replace an old large electric appliance such as a heater, clothes dryer, hot water heater, stove, or oven, they can reap significant energy savings and lower costs by choosing a more efficient gas appliance. In general, replacing older appliances, with newer ones also has big energy advantages. A new, high-efficiency gas heating system uses half the natural gas or oil as a 20-year-old one. New Energy Star refrigerators can use 50% less electricity than those that were manufactured 15 years ago. Compact fluorescent lights (CFLs) use up to 75% less electricity than conventional incandescent light bulbs while producing the same amount of light. Air conditioners are 20-50% more efficient than they were 15 years ago. Also, air conditioning systems can be made 10-20% more efficient if homeowners would just repair leaky ducts and correct refrigerant levels. Businesses can benefit even more from taking similar actions.

But we have to act collectively to have a true impact and bring down costs. Becoming more energy efficient is one of the cheapest ways we can start regaining energy independence. Acting together in large numbers will drive down costs, help us gain momentum and further accelerate the pace of change.

A Relentless PR Campaign

What often prevents us from acting wisely is a failure to track our expenses. We fail to recognize that the higher price we pay for an energy efficient appliance is compensated for by the money we save from its lower cost to operate. Customers often don't figure in the cost savings over time. Publicizing the cost benefits we get from greater energy efficiency can help. But publicity shouldn't stop there.

Publicity is one of the greatest promoters of change in our society. It has been extremely successful in overcoming the natural inertia that prompts us to keep things as they are. Therefore, I add publicity as a major component of my plan. The government needs to wage a relentless public relations (PR) campaign that promotes alternatives to hydrocarbons. Through publicity, it can change perceptions, gain public support, and motivate action that creates momentum for change.

To work, the PR campaign must make it simple, socially fashionable, and cost effective to save energy through our purchases. Beyond that, the government needs to stress energy self-reliance. The government was able to wage a successful anti-smoking campaign. It should be able to wage a successful campaign for alternative energy and energy efficient appliances?[8]

United Alternative Energy Advocacy Group

To help win the public over to alternative energy and greater energy efficiency, the various alternative energy advocacy groups need to unite in a single coalition across the country, even as they continue to campaign for their preferred energy alternative. A unified organization of advocacy groups will make a greater impression on the public and the press and could dedicate more resources to publicity that increases the public's awareness of alternatives. They can explain why we must change and what is available. The government can't succeed alone. It needs help from advocacy groups and the press. We need unity of purpose for the greater good.

Incentives and Legislation

Complementing these actions, I add to the plan's component list continued and extended economic incentives in the wider community outside the alternative energy zones. This carrot approach provides rebates or subsidies that level the playing field for alternatives, increasing their competitive advantage. We must further narrow the gap in price between cars that meet high mpg and clean air standards and those that don't and between devices and appliances that merit Energy Star support and those that don't. One key is to always give consumers a choice, but make it easier, more desirable, and increasingly more economical to choose what is energy-efficient.

For this approach to work, we need more energy efficient appliances, more hybrid cars, etc. Increasing incentives is instrumental to encouraging increased production. American car companies, for example, are more likely to produce fuel efficient hybrids and other fuel efficient vehicles if they can offer these cars at competitive prices. American car companies like any other business are in business to make money. They have a natural concern about getting too far out ahead of the public. Their strategy of emphasizing hybrid SUVs rather than more fuel efficient hybrid sedans may reflect this concern and a desire to emphasize their core strengths. They evidently feel that the public is not ready to move to the smaller hybrids. On the other hand, American car manufacturers have to be concerned about lagging behind their European and Asian competitors in perfecting smaller, more fuel efficient cars if these cars increase in popularity as they did in the 1970s. Economic support and legislation can encourage American car manufacturers to decide in favor of the smaller hybrid cars.

While the demand exceeds supply, the federal government must also move to stop car dealers from unreasonably profiteering from supply shortages, which will discourage buying energy efficient vehicles.

Use Waste to Reduce Waste

Recycling is another component I would add to the list. We can reduce hydrocarbon use more quickly by increasing our capability to recycle waste products. By using recycled and refurbished products, we save energy as well as manufacturing costs. Entrepreneurs should be encouraged to establish recycling where it doesn't already exist and expand it where it already does exist. Also in the waste management area, as part of an alternative energy program, state governments should experiment with microbial fuel cells and other new technologies with multiple-uses.

The ability of microbial fuel cells to clean up waste water, while producing hydrogen, ethanol, methanol, and other renewable fuels makes them more potentially profitable. We should seek to create multi-purpose alternative energy plants to increase profitability. Making alternatives profitable is a key to reducing government costs of transition and to stimulating the market to act instead.

The Coal Card

We are currently being squeezed between our need for oil to prevent a potential economic and social disaster and our need to substitute alternatives to avoid a potential environmental disaster. Even with breakthroughs, we have to make better use of the hydrocarbon resources we have – so I add the judicious use of coal to my components list. As I've already argued, we can use less polluting gasified coal to produce more electricity and hydrogen for fuel cells and in the manufacture of more efficient appliances and alternative energy conversion devices. Though it is a less efficient use of coal, we can also promote coal conversion to other fuels (oil, ethanol, hydrogen) to buy us time to transition to other alternatives, and we can recycle CO_2 emissions from coal to increase the yield of alternative fuels.

I would still exclude further development of nuclear power because of its long lasting waste products, but recognize the enormous pressure we will have to use it too if we fail to mitigate our energy risks using cleaner alternative energy.

A Price Floor

Another vital component of the plan is a price floor for oil, which combats price fluctuations. We must prevent ourselves from reverting to wasteful habits as we did in the 1980s, whenever prices for oil fall significantly.[9] Looking back, it should be clear that our failure to put more effort into transitioning to renewable alternative energy in the 1980s and 1990s has cost us dearly. Now that we are in a situation that is far more precarious, we can ill afford to make the same mistake again. We cannot let those who would keep us dependent on oil veil our vulnerability beneath lower prices and propagandistic obfuscation.

To prevent us from backsliding and stalling, we need to allow gasoline to rise closer to its true price; we need a price that is at least high enough to give alternatives a chance to compete and which stimulates us to be less wasteful.

While I think it is extremely unfortunate that we have made the price of gasoline such a driver, I'd rather see price used effectively to compel action than have us slide back into old habits. I would suggest a threshold of $90.00 per barrel in 2008 dollars, with adjustments for inflation. Our economy has shown we can tolerate that price given that we have functioned with a price per barrel above $100.00.

If we finally get a grip on the conventional energy supply by reducing our need for it and if prices subsequently decrease, we can use the extra money that a maintained floor price will provide to fund the alternatives and further accelerate the pace of change.

Countermeasures and Contingency Plans

At the same time, we have to prepare for the possibility that oil producing countries will manipulate oil supplies to thwart attempts to substitute alternatives. That is one reason why I add contingency plans to the list. We need to reduce the amount of imported oil we consume by diversifying our energy base. We have to broaden our energy base before we are forced to pay so much for oil that we can't afford alternatives. But the energy producing countries will almost certainly push back. It shouldn't be surprising to find that any attempt to reduce oil consumption, regardless of the reason, will meet with opposition from those who don't want to see us make progress toward energy independence at their expense. Price manipulation and supply disruptions are their possible weapons of choice to prevent real progress from being made. We need to prepare countermeasures and contingency plans to reduce the threat of energy blackmail from whatever quarter it comes.

As competition for oil heats up, agreements that reduce our supply of imported oil are going to be signed anyway, quite independently of any efforts we make or don't make to diversify our energy resources and reduce our oil consumption. So we might as well face the threat of supply cuts head on. While we need to transition with as little impact to the economy as possible, we can't allow ourselves to be intimidated into continuing the status quo. The status quo is not really a status quo; it is an inexorable, though still relatively slow, slide into economic disaster. Remember what even some of our own companies did to extract money from California in 2000 with just partial monopolistic control of grid energy resources. Now apply that malevolent creativity to the oil cartel. Given what we face, we must include real contingency plans to combat fluctuations in price. We can't take the complacent attitude that the strategic oil reserves will bail us out. At the very least, all states from the top down to the community level need to make plans for:

- Expanding Carpool networks
- Beefing up the capability of employees to telecommute
- Increasing use of public transportation

- Instituting a gasoline rationing system
- Encouraging carpooling
- Distributing the pain of shortages equitably[10]
- Instituting incentive-driven voluntary reductions of energy use in times of shortage

Carpooling and telecommuting should be encouraged in general, not just as part of a contingency plan. Obviously, more will be necessary. Yet at this writing, I haven't found a single state or federal contingency plan that goes far enough toward an integrated, well thought out, coordinated plan. Much of what I'm advocating is common sense.

A Phased Approach

The components of this plan should play out in three phases.

1. The Bridging Phase

During the first phase we should be emphasizing our best near term capabilities, many of them bridging technologies such as the hybrid car. We should be increasing public awareness of what's at stake, increasing the spread of energy efficient appliances, and creating contingency plans, some of which we can implement right now, such as promoting telecommuting at least one day a week. We should take advantage of our most abundant hydrocarbon, coal, to build and transition to alternatives. And we should consider recapturing CO and CO_2 emissions from polluting power plants to make alternative fuels. We should increase the use of alternative fuels including hydrogen across the country. We should exploit existing infrastructure to connect the transportation and stationary power sectors following the guiding concept I've mentioned. And we should be establishing an ever increasing number of alternative energy zones to focus and refine our efforts.

Within the alternative energy zones, we should create a real-life laboratory where an entire alternative energy infrastructure can be evolved and perfected – incubating the future. With the help of customers eager and dedicated to "going green," we can test the most promising advances from nanorod solar cells to hybrid fuel-cell/gas-turbines. We can build public confidence in alternatives, extend their use, and lower their costs.

During this phase we should establish a floor price for oil, and use the taxes that maintain a floor price to subsidize alternatives to oil and coal. Furthermore, as alternatives spread and pollution levels fall, we should develop the capability to measure the drop in pollution by region. By the end of this phase, we should have reduced oil imports by 50% and reduced manmade pollution from hydrocarbons by 40% from today's level.

2. The Breakout Phase

During phase two the alternative energy economy and its integrated, distributed infrastructure should evolve from and supplant the existing infrastructure. It should interconnect communities, spreading out from established local (that is, distributed) alternative energy economic zones. This is the phase too in which an extensive alternative energy transportation network can be built.

The bridging technologies may be expected to have evolved during this phase to more advanced models. Hybrids may be replaced, for example, by fully electric cars that use advanced energy storage technologies, fuel cells, advanced electric motors, and other innovations. Electricity should increasingly tie the stationary power sector to the transportation sector. And pollution levels should continue to fall. We should have reduced our national dependency on foreign imported oil to less than 15% of our needs by the end of this phase and reduced manmade pollution from hydrocarbons to 30% of what it is today.

3. The Consolidation Phase

During the third phase the alternative energy infrastructure should become well established, extending across the country. Cars should become part of the stationary power structure, receiving and contributing electricity to the home and business as we achieve the full integration of the stationary power and transportation sectors.

The complicated electrical grid that crisscrosses the country should become a backup system for much of the country, needed mainly to achieve equitable distribution of electricity to places less able to achieve complete local energy independence. And we should no longer need to import foreign oil or natural gas. Manmade pollution from hydrocarbons should be less than 20% of what it is today.

The Real Thing

The ultimate challenge in the next twenty years will be to make alternative energy a real alternative. What does that mean? The easiest way to make the transition is to provide people with what they are familiar with and evolve from there. Simply put, green cars must look and handle like conventional cars, have at a minimum a 250-300 mile range on a "tank" of fuel. And they must have a good safety record. Electricity generated from renewable alternative energy sources should run household electricity appliances just as effectively and ultimately more cost effectively than electricity generated from hydrocarbons.

Alternative energy in the stationary power sector will have relatively high up front costs, but should have low sustaining costs. Electricity generated from fuel cells, wind, and solar energy should pay back their owners in energy cost savings in 7-10 years, given what customers would have paid for conventional electricity.

The next two chapters will describe how we could systematically go about expanding alternative energy using the components of my plan, the market, and the government in synergy.

26

The Near-Term Campaign

To implement the plan I proposed in chapter 25 requires a carefully thought out strategy that follows the **guiding concept** of integrating the stationary power and transportation sectors using electricity. I've divided the strategy into two parts: a near-term strategy described in this chapter and a longer-term strategy described in the next chapter. The near-term strategy is outlined in this process diagram (Figure 34).

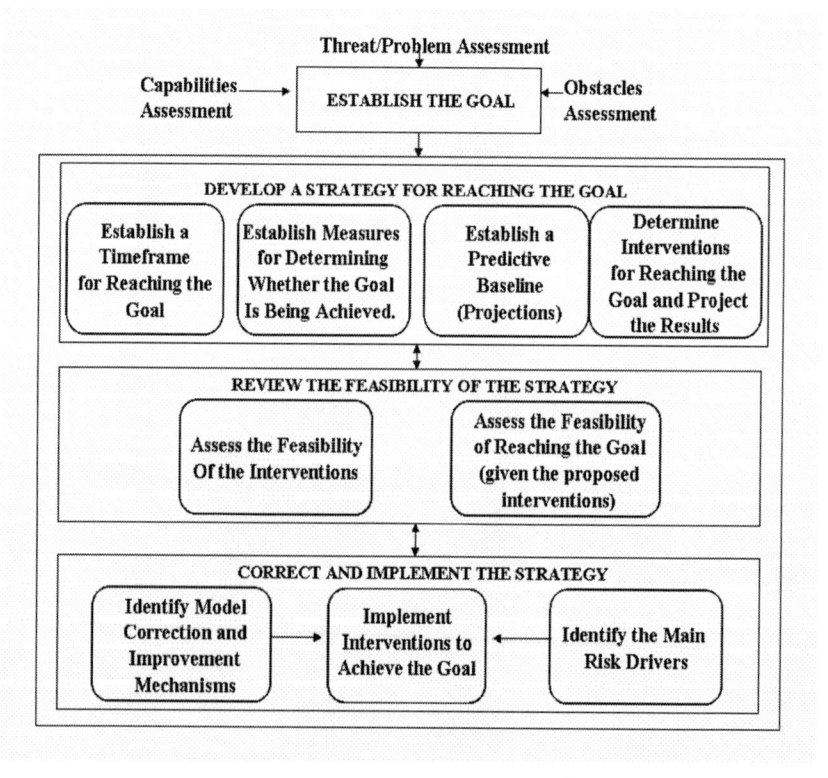

Figure 34. Process Diagram[1]

A process diagram is simply a tool for thinking through how to systematically do something. This process diagram is hierarchical and starts with a goal. The goal is determined by assessing what the threat or problem is, what capabilities exist to solve it, and what impediments to a solution exist. These three considerations were covered in preceding chapters.

Once the goal is determined, the process calls for the establishment of a strategy for reaching the goal. The feasibility of the strategy and goal are considered. If the strategy appears workable and the goal appears achievable, the strategy is implemented. The process then uses feedback to check the results of carrying out the strategy and makes corrections or improvements to the process as needed. The rest of this chapter will discuss each part of the process diagram in Figure 34 except actual implementation. It starts with the goal.

Establish the Goal

The goal is to reduce the amount of oil imported into the U.S. by 50% from what it is today, despite an expected growth in population. We consume 20.8 million barrels per day (mb/d) today and since 13.7 mb/d is imported, the goal is to reduce imported oil to 6.9 mb/d. This goal should be met in such a way that it abides by the guiding concept and ensures that pollution levels are reduced at the same time by at least 40%.

Establish a Timeframe for Reaching the Goal

The timeframe for achieving the goal is twenty years.

Establish Measures for Determining Whether the Goal is Being Achieved

The amount of oil imported into the U.S. each year is the main measure of whether the goal is being met.

Establish a Predictive Baseline

The baseline projections predict what we would likely face in twenty years if current trends in energy consumption were allowed to continue.

Baseline Projections for Twenty Years from Now

- U.S. petroleum consumption will have increased by 32% to 27.5 million barrels per day (mb/d).
- Imported oil will have risen to 22.6 mb/d from 13.7 mb/d.
- Domestic oil production will have declined slightly to 4.9 mb/d
- Of all the oil used, 64% will go to produce gasoline and diesel fuel; that is 17.6 mb/d.
- Electricity consumption will have increased by 63% from 4.1 to 6.7 billion megawatt-hours per year.
- The average light duty vehicle (passenger car, SUV, pickup truck) will get only 21 miles per gallon.

These projections are based on historical trends but are not an exact reflection. The statistics that were used are from the Department of Energy's Energy Information Agency (EIA), the media, oil companies, noted institutes, think tanks, and some websites that provide statistics related to oil.[2] The statistics from different websites weren't entirely consistent[3] so the degree of their mutual consistency was used as an indication of the reliability of the data. Greater weight was given to government statistics. Table 4 (on the next page) shows the trends over a twenty year period between 1985 and 2005.

Table 4 indicates that U.S. petroleum consumption increased by 33% between 1985 and 2005. So the baseline projection was chosen to continue this trend with another 32% increase in the next twenty years (starting 2008). The projected increase in foreign oil imports from 13.7 to 22.6 mb/d is much larger than the 43% increase between 1985 and 2005, The higher number for imports assumes domestic production will not rise above what it is now; in fact it will decline slightly, but oil imports will steadily increase, which widens the percentage gap between them over time. The projection takes into account recent finds in the Gulf of Mexico and more efficient methods of extracting oil from well-known, otherwise played out domestic oil fields like that in the Bakken Basin. If domestic oil production can not be maintained close to current levels, and we do nothing else to reduce our dependency, we will slide into even deeper dependency. The baseline projection for the amount of oil expected to be used for gasoline and diesel fuel is based on 1985 and 2005 statistics.[4] These statistics indicate that 60-64% of the oil produced

THE NEAR-TERM CAMPAIGN   255

was refined into gasoline and diesel fuel. The percentage used in the baseline projection is 64%.

Table 4. Historical Record – Main Findings[5]

Basic Information	1985	2005	1985 to 2005
World Population (billions)	4.8	6.4	33% increase
U.S. Population (millions)	230	295	28% increase
U.S. Petroleum Consumption (millions of barrels per day)	15.7	20.7	32% Increase
Oil Production in U.S. (millions of barrels per day)	8.9	5.1	43% decline
U.S. Proved Reserves of Crude Oil (millions of barrels)	28,446	21,757	24% decline
Net Petroleum Imports (millions of barrels per day)	6.7	12.4	85% increase
Net Petroleum Imports (% of U.S. Petroleum Consumption)	42	60	43% increase
U.S. Electric Power (Net Generation) (billions of kilowatt hours)	2,473	4,038	63% increase
U.S. Electric Power Consumption (Retail Sales) (billions of kilowatt hours)	2,473	4,055	64% increase
Number of Passenger Cars U.S. (millions)	128	137	7% increase
Number of "Light Trucks" (SUVs, Pickup Trucks) (millions)	37	95	157% increase
New Passenger Cars U.S. (% per year of passenger cars)	2%	2%	0% increase
Light Duty Vehicles (Passenger Cars, SUVs, Pickup Trucks) (mpg)*	21.8	20.2	7% decrease

*Adjusted based upon 2008 ratings

The baseline projection of a 64% increase in electric power consumption is based on a 63% increase between 1985 and 2005. The projection that car fuel efficiency would remain about the same as it is now at 21 mpg, is based on the period between 1985 and 2005 which ended up with a 7% decrease in the average mpg for light duty vehicles. The last two years there has been a slight increase in average mpg.

The baseline projections reflect the assumption that we don't change our consumption habits easily. Our habitual inertia is reflected in a trend toward greater consumption (of all sorts) per person. Because increases in energy consumption are being balanced by current increases in the energy efficiency of the gadgets we use, population growth has become the dominant driver. This view is supported by the historical trends. If you look at population growth in Table 4, you'll see it correlates with the increases in petroleum consumption between 1985 and 2005.

Determine Interventions for Reaching the Goal

The term "interventions" in the process diagram (Figure 34), refers to actions that can be taken to change current trends in order to reach the goal. The key interventions, which are listed below, utilize the plan components described in chapter 25.

- Produce additional domestic oil by coal-to-oil conversion
- Increase the number of flexible fuel, plug-in hybrid and electric cars, and increase the use of alternative fuels (e.g. ethanol and hydrogen)
- Increase recycling and improve waste management
- Increase the use of telecommuting and carpooling
- Establish alternative energy zones
- Increase the number of energy efficient appliances and electronic devices on the market
- Promote insulation, passive solar and related technology in homes and businesses for heating and cooling
- Use computer controlled energy systems to automate energy efficiency.
- Promote renewable solar, wind, ocean wave, and fuel-cell/gas-turbine generators

In addition, there are political and social interventions that will increase the chances of reaching the goal. They include: an increase in publicity about alternative energy and the need for change; a ban on TV advertising of gas guzzlers; an increase in incentives and subsidies for those who produce and buy green products; the use of system grants in

THE NEAR-TERM CAMPAIGN

alternative energy zones; enactment of more green legislation in support of the goal; and the establishment of a floor price for oil.

Project the Results of the Interventions

To reach the goal of reducing oil imports from what they are today by 50% (and emissions by 40%), the first 5 interventions listed need to achieve a number of results indicated here:

Results of the Interventions Twenty Years from Now

- Domestic coal-to-oil conversion will have decreased the need for imported oil by 6.4 mb/d from the baseline projection of 22.6 mb/d of imported oil.
- Light duty vehicles will get an average of 51 miles per gallon, which decreases the amount of imported oil required by 7.3 mb/d.
- Greater recycling of usable products and metals (which saves energy) and stepped up production of biofuels, syngas, and hydrogen will have reduced the need for imported oil by 1.6 mb/d.
- Telecommuting at least one day a week and/or a four day work week for one third of the people with cars will have reduced the need for imported oil by 0.3 mb/d.
- A sufficient number of alternative energy zones will have been set up throughout the country to reduce the need for oil imports by 0.1 mb/d.

Shifting energy consumption from the transportation sector to the stationary power sector is likely to increase stationary power consumption by as much as 100% from what it is today, to 8.2 billion MWh per year with 37% due to plug-ins. This anticipated increase in stationary power consumption can be offset if we could achieve the following desired results using the remaining interventions on the list.

- Improved efficiency (e.g. more energy efficient appliances, the use of passive solar, solar thermal, better insulation, software energy control) will have reduced the 8.2 billion MWh per year consumption of stationary power by 45% to 3.7 billion MWh per year.[6]

- Renewable energy such as wind and solar, along with coal gasification, syngas, and renewable biofuels, will supply at least an extra 0.4 billion megawatt-hours per year.

Assess Feasibility of the Interventions

To assess whether the goal can be met, the feasibility of reaching the desired results will be considered for:

Interventions
1. Coal-to oil conversion
2. Flexible fuel, plug-in hybrid and electric cars
3. Recycling and waste management
4. Telecommuting and carpooling
5. Alternative energy zone
6. Improvements in home and business energy efficiency
7. Renewable energy turbine generators

1. Coal-to-Oil Conversion

How feasible is it for a domestic synthetic oil industry to produce 6.4 mb/d twenty years from now? The coal gasification process for making oil certainly exists. It was discussed in Chapter 13.[7] But it would be expensive to build the processing plants that are needed. According to one report, it would take $10-11 billion to build a U.S. plant that produces a hundred thousand barrels of synthetic oil a day.[8] To produce 6.4 mb/d we would need at least 64 such plants[9] at a projected total cost of $640-704 billion over 20 years.

What about the feasibility of building 64 plants in 20 years? It is a question of economics rather than building capability. Based on the analogous situation of building coal burning electrical plants, we certainly should have the capability to build 64 coal-to-oil conversion plants in 20 years.[10] China has already demonstrated an analogous capability. It is building coal fire power plants at a much faster rate.[11]

For companies to invest in the plants, however, there needs to be some reasonable expectation of a return on investment. Is coal-to-oil conversion profitable? The price of imported oil exceeded $98 a barrel in November 2007 and over $100 a barrel in early 2008. Government estimates of the cost to produce a barrel of oil from coal range from $40

2. Flexible Fuel, Plug-in Hybrid and Electric Cars

to $60, including the cost of CO_2 sequestration for zero emissions. Some estimates are even lower.[12] If government estimates are accurate, oil from coal would quickly become profitable for companies.[13]

What about pollution? Coal gasification with sequestration is expected to emit close to zero air emissions. Nonetheless, standards need to be set to ensure pollution of any sort is at least 60% below the current emissions of coal fire power plants.

Regarding vehicle fuel efficiency increases, how feasible is it for the average light duty vehicle to get 51 mpg? There is little doubt that we could create commercially available cars that get 51 miles per gallon in 20 years, given what is already possible. The Honda Civic Hybrid gets around 40 mpg and the Toyota Prius hybrid gets 47 mpg, based on the new and more stringent 2008 rating system. Prius owners who have modified their cars to take supplemental batteries and to plug into a home electrical outlet for recharging are reporting 100 mpg.[14] Both the Prius and the Civic Hybrid, in addition to their superior fuel efficiency, emit 90% less pollution than conventional cars on the road today, which indicates that cars like them can be key to helping us reduce overall emissions by 40%. Looking toward the future, there are bound to be even greater improvements in electric motors, fuel cells, batteries, and other energy storage devices over twenty years if we make fuel efficiency a high national priority. But can we produce enough fuel efficient, commercially viable cars to achieve the desired result in 20 years? That is where the challenge may lie. In Table 5 below, I suggest a 20-year scenario that would achieve the situation in which light duty vehicles (passenger cars, SUVs, and pickup trucks) average 51 mpg.

The table assumes a ramp up strategy in which car manufacturers steadily switch over production to hybrids, plug-in hybrids, and other fuel efficient cars while steadily reducing production of gas guzzlers. In the Table 5 scenario, by the end of the tenth year the newer light duty vehicles would constitute 5% of light duty vehicles on the road and would average 60 mpg.

The older light duty vehicles would average 25 mpg. In the fifteenth year, the newer light duty vehicles would constitute 9% of the vehicles on the road and would average 80 mpg. The older light duty vehicles would get 33 mpg. The rise in the mpg average for the lower mpg cars would be due to older, less fuel efficient cars being progressively junked, leaving the "older cars" with progressively higher mpg. By the twentieth year, 10% of the light duty vehicles on the road would be "newer cars" and would average 100 mpg. The rest of the light duty vehicles would average 46 mpg. The overall average would then be 51 mpg.

Table 5. Strategy to Increase MPG of Light Duty Vehicles[15]

Year	% Newer Light Duty Vehicles	Ave MPG	% Older Light Duty Vehicles	Ave MPG	Overall Ave MPG
Ramp Up in 10 years	5%	60	95%	25	27
15	9%	80	91%	33	37
20	10%	100	90%	46	51

[At a particular time in Table 5 (i.e. 10, 15, or 20 years), "newer" refers to cars less than 5 years old and "older" refers to cars 5 years old or older.]

Why be confident that such an increase in fuel efficiency is feasible? Besides the fact that we already have commercially available, fuel efficient cars, we have done even better before. By 1974, after the oil embargo, American car manufacturers realized they would have to produce smaller, more fuel efficient cars to compete with the Japanese. And Congress realized we were in economic trouble. Both car manufacturers and Congress took action. Congress mandated higher mpg. The American car companies reduced the average size of their cars and increased their mpg. It took about 12 years to become competitive, but the results were impressive.[16] Between 1973 and 1985 the average mpg for passenger cars went from 13 to 21.8 mpg.

But Congress provided a loophole for car companies to make as many gas guzzlers as they could sell by allowing light trucks (SUVs and pick up trucks)[17] to meet a much lower mpg requirement.[18] In 1985 when the price of a barrel of oil began to plunge, car manufacturers promoted profitable, gas guzzling SUVs for use as passenger cars. Not surprisingly SUV sales began to soar. Between 1985 and 2005 the number of SUVs on the road rose 157%. Meanwhile, the average mpg for light duty vehicles dropped by 7%. In effect we had retreated from the efficiency standards that had been set.

What Table 5 is meant to suggest is not only the difficulty but also the feasibility of making the transition to more fuel efficient cars. It can't happen overnight, but it can happen over 20 years if government policy and customer pressure mandate progressively higher fuel efficiency year by year.

The best way to reach 51 mpg in 20 years is to lower the price of fuel efficient cars and to make them popular. If we use a portion of gasoline taxes to give rebates for the purchase of high mpg cars, especially hybrid plug-ins, we can greatly encourage the purchase of hybrids. In order to reduce the price of high mpg cars, it will also be necessary to increase production. My plan calls for state and federal governments to pass legislation that encourages the production of fuel efficient cars. Following the guiding concept, plug-in vehicles should be favored. But the increased use of hybrid plug-ins will also require the state and federal government to take measures to increase the production of electricity for the plug-ins.

To promote sales of hybrids further, the floor price for oil should be set to a price that allows alternatives to compete. So it will be useless for producing countries to dump oil onto the market to lower the price of gasoline so that gas guzzlers are more tolerable and there is less motivation to buy hybrids. In addition, states should ban enticing ads for gas guzzling cars as cigarette ads were once banned. Publicity at the state and community level will be required to explain why these actions are needed.

If light duty vehicles will be averaging 51 mpg in twenty years rather than the 21 mpg of today, then we would require only 41% as much gasoline and diesel as we would in the baseline projection. That percentage decrease in gasoline and diesel corresponds to a reduction of 7.3 mb/d from the projected baseline of 22.6 mb/d for imported oil.[19] This estimate is based on a previous assumption that light duty vehicles would constitute 70% of all vehicles on the road.

3. Recycling and Waste Management

Next consider diversifying our fuel base to reduce the need for oil. Some might point to Brazil as a model for success. Brazil is able to produce 0.33 million barrels of ethanol a day,[20] which has had a huge positive impact on their economy. But Brazil is able to produce ethanol from sugar cane while we rely mainly on corn, which yields less ethanol than sugar cane does.[21] And there are other issues already covered, which suggest that diversifying our fuel base using corn or even sugar cane is not a good option for most states. But there are other options. Our land fills and waste (sewage) water hold significant amounts of untapped energy. Using some of this untapped resource and recycling usable products and metals to save energy, Waste Management Inc., the largest waste management company in North America,[22] estimates that during the last 20 years it has reduced oil imports by 5 mb/d from what they would have been.[23] Given what this one company has been able to accomplish, it seems reasonable to expect that we can reduce our oil dependency by an additional 1.6 mb/d in the next twenty years if we expanded on what Waste Management Inc. is already doing. Expansion is possible because we have tapped less than 1/3 of garbage, sewage, other waste biomass, and recyclables [24] that are available today and because more is always being created.

4. Telecommunicating and Carpooling

What about changes in workforce patterns? How feasible is it to expect that enough of the workforce will telecommute or carpool at least one day a week or choose a ten-hour day four-day workweek? For one thing, some companies already offer telecommuting, carpools, and four day forty hour workweeks. The Internet makes telecommuting within reach of most companies. Companies or services are currently available that match prospective carpoolers with one another based on the location of home and work. On the other hand, there are jobs today and presumably twenty years from now that couldn't be done on a four day, 40 hour work schedule. Examples range from harvesting crops to shooting a film. Then there are questions about Internet security and employer fears of lost productivity if they allowed workers to work from home. Other objections may come from employees. There are employees who would refuse to car pool, for instance. But none of these concerns seems to me to be a showstopper.

The significant increase in the number of cars crowding our freeways due to population growth should certainly encourage telecommuting and carpooling. More states could institute incentive programs to reduce the number of cars on the road per day with the benefit that those states would require less money for freeway expansion. Telecommuting one day a week or more or working four days a week 10 hours a day should become very popular. Companies may even extend telecommuting to the entire week, making working from home the norm for many. Carpooling may not be as popular, but if oil prices continue to be high or even climb as expected and congestion increases as expected, workers will do the rest. If one third of people with cars don't use their cars just one workday a week, about 5% less gasoline would be used, which amounts to 0.3 mb/d if the average car gets 51 mpg.[25]

5. Alternative Energy Zone

As for the alternative energy zones, could they reduce oil imports by 0.1 mb/d? Unfortunately, it is hard to find modern precedents with which to determine whether this result could be achieved. But it is possible to estimate what it would take, given one of several possibilities.

I would start by assuming that the alternative energy zones twenty years from now are oil free; that is, they should have cars that get all their energy from electricity and homes that get all their electricity from alternative energy. Is that possible? Since electric cars already exist, and some homes can already obtain all their power needs from alternatives, it should be possible for an alternative energy zone to run entirely on electricity 20 years from now. But being oil free is only part of the consideration. If we established 300 alternative energy zones throughout the country in 20 years, each alternative energy zone would have to reduce energy consumption by an amount equal to the energy from 333 barrels of oil a day to reduce oil imports by 0.1 mb/d.

Today, the average household consumes the equivalent of 1/4 barrel of oil per day in the form of petroleum products from plastics to gasoline,[26] while an average small business consumes roughly the equivalent of 1 barrel of oil per day.[27] Using these rough figures, the equivalent of 888 homes and 111 small businesses would have to participate in each alternative energy zone to reduce oil imports by 0.1 mb/d.[28] A reduction of 0.1 mb/d is not much compared to that of other near-term interventions but that is because zones are the seed for a longer term solution. The contribution of alternative energy zones to reducing oil imports will grow in importance over time. Chapter 27 defines their long term role in ending our oil dependency.

6. Improvements in home and business energy efficiency

Regarding home appliances, how feasible is it to improve home efficiency by 45% in 20 years? If we replaced all major appliances that are 15 years old or older and all major electronic devices that are five years old or older with the most energy efficient equivalents on the market, we would save from 30-75% per item. Here is a partial breakdown of the percentage savings in energy consumption for some major household items if old ones were replaced:

- Copiers and Fax Machines   20-25%
- TV/VCR/DVD   20-30%
- Air Conditioning   20-50%
- Clothes Dryer/Washer   30-40%
- Dish Washer   40-50%
- Electric Range Oven   30-70%
- Refrigerator   35-50%
- Water Heater   40-70%
- Computer with Monitor   50-60%
- Lights   60-75%

The range of percentages is based on the Energy Information Administration and the Federal Government's Energy Star websites, which provide specific percentages based on size, capacities, and capabilities of the various appliances and electronic devices.[29]

To put these savings in context, another bit of information is needed. Figure 35 indicates what percentage of the average household's total energy usage is consumed by various appliances and electronic devices according to Energy Star.[30]

Pie Chart of Energy Consumption

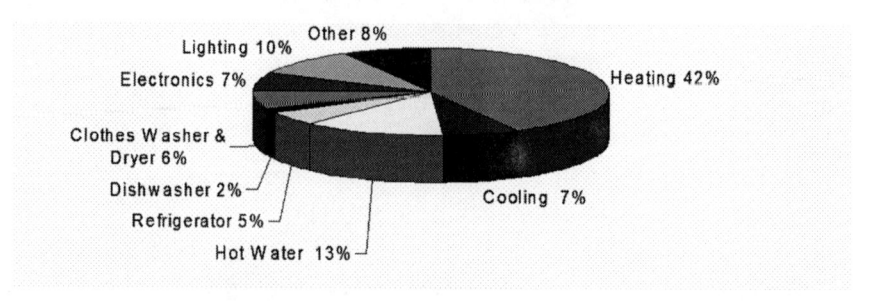

Lighting 10%
Other 8%
Electronics 7%
Clothes Washer & Dryer 6%
Dishwasher 2%
Refrigerator 5%
Hot Water 13%
Cooling 7%
Heating 42%

Figure 35. Home Energy Consumption[31]

THE NEAR-TERM CAMPAIGN

Bosgraaf Homes which builds Energy Star homes claims that their homes are 40-50% more efficient than homes built to the 1993 National Model Energy Code.[32] The Department of Energy's new mandate for 2008 provides a further suggestion of where we are. The DOE is requiring all new federal buildings built in 2008 (which includes federal housing) to be 30% more efficient than prevailing building codes require.[33]

Given these statistics, we should be able to increase the average energy efficiency per household and business by at least 25% in 20 years if energy inefficient appliances and electronics are replaced by energy efficient ones. We should be able to add at least another 20%, if homeowners and businesses would optimally insulate their premises, use passive solar heating and thermal solar, and change habits that waste electricity.[34] Already there are devices that could help us automatically save electricity by shutting off lights when no one is in the room, for instance. And more is possible, such as smart computer and sensor control of energy consumption to avoid waste. The possibility of their widespread use 20 years from now is enhanced by the fact that such systems already exist.[35]

Legislation that favors green products and taxes energy guzzlers could help change behavior. Local and state governments, advocacy groups, and the media could make it acceptable to use tax money from energy guzzlers to reduce the price of energy efficient appliances.

7. Renewable Energy Turbine Generators

What about generating an additional 0.4 billion megawatt-hours per year in the stationary power sector using wind and solar, along with coal gasification, syngas, and renewable biofuels? With alternative energy – not considering hydroelectric power or nuclear power – comprising only about 3% of energy used to generate electricity, that result may seem impossible to achieve. But the necessary power generating technology already exists even without breakthroughs. And the use of alternative energy is beginning to rise.

The main problems are cost and a negative image. To succeed, the price of alternatives must come down and confidence in them must increase. States will also need to develop the smart grid technology necessary to integrate community based alternatives into the grid system.

Assuming prices for alternatives become reasonable, we should be able to produce an additional 0.4 billion megawatt-hours per year if, we increase installed capability so that 10% of all power consumed above current levels came from renewable sources such as wind, solar, wave, hydrogen, geothermal or a combination of renewable energy sources[36] This increase should be possible given current trends,[37] and like increases in energy efficiency is measurable year-by-year.

We can increase our chances of succeeding if we take greater advantage of off-peak hours at night and in the wee hours of the morning.[38] A DOE study suggests that 84% of cars today (if they were plug-ins) could take advantage of off-peak hours to recharge without increasing the need for new power plants.[39]

In short, we should be able to draw down the baseline projected increase due to normal energy consumption (which excludes vehicle plug-in consumption) from 6.7 to 4.5 billion MWh per year if we can become more efficient in our use of energy at home and at the workplace. And we could counter an increase in power consumption due to vehicle plug-ins by exploiting off-peak hours use and through a 0.4 billion megawatt-hours per year increase in capacity generated from renewables.

Furthermore, we could supplement wind, solar, ocean wave, and geothermal generated energy with electricity generated using coal gasification, syngas, and biofuels.

Assess the Feasibility of Reaching the Goal

To assess the feasibility of reaching the goal, I consider whether the goal can be reached if the interventions achieve their desired results. According to the baseline, if we continued current trends, if shortages of oil and natural gas did not materialize, and if no catastrophic events occurred in twenty years:

- We would consume 27.5 million barrels of oil per day (mb/d)
- We would produce 4.9 mb/d of oil domestically.
- We would require 22.6 mb/d from foreign countries.
- Of the 27.5 mb/d consumed, 17.6 mb/d goes to gasoline and diesel

Now here again is the amount we could subtract from the 22.6 mb/d of imported oil if we achieved the desired results.

THE NEAR-TERM CAMPAIGN

- Coal to Oil conversion: 6.4 mb/d
- Fuel Efficient Cars: 7.3 mb/d [40]
- Biofuel Production and Recycling: 1.6 mb/d
- Telecommuting: 0.3 mb/d [41]
- Alternative Energy Zones: 0.1 mb/d

If we take into account all these reductions, the total amount of imported oil would drop to 6.9 mb/d or 50% of what it is today. Note the key near-term drivers in the success of this strategy are the increase in fuel economy and increases in efficiency of homes and businesses.[42]

To improve the chances of reaching the goal, the government should step up research in specific targeted areas where improvements could have the greatest short-term impact. The areas that would be most important given the guiding concept are:

1. Plug-in Hybrids and Electric Cars. These are key to integrating the transportation and stationary power sectors. Research should aim at increasing their efficiency and bringing down their price so consumers will buy them. Part of the research can also aim at increasing the efficiency of the internal combustion engines in gas-electric hybrids by capturing heat from combustion to produce electricity.

2. Energy Storage Devices. Energy storage is instrumental to both the stationary power and transportation sectors if alternative energy is to be viable. Research should aim at reducing costs, increasing capacity, increasing durability, and ensuring their safety.

3. Solar Thermal Devices and Passive Solar. The heating and cooling systems in homes, businesses, and public buildings consume a large portion of stationary power. Solar thermal devices could be used both to generate electricity and provide heating and cooling. Passive solar can be used for heating and cooling. Research should aim at exploiting their low cost potential and potentially combining passive solar and solar thermal systems.

4. Coal Gasification. In the short term we will probably have to increase our use of coal. Research should aim at developing cleaner

coal gasification techniques to reduce environmental damage and increase efficiency. We should aim to reduce emissions by 60% below current levels. As part of their work, researchers should consider the impact of carbon dioxide sequestration versus carbon dioxide capture for use in producing oil and gasoline substitutes.

5. Power Grid. The grid is essential for evolving a new energy infrastructure that ties the stationary power and transportation sectors together. Research should aim at creating smart grids that can integrate renewable energy smoothly into its infrastructure. Research should look into the effectiveness of storing renewable energy for later grid use. (Energy storage for use by the grid is not currently done.) And grid research should look into evolving the grid into a distributive (local) system to increase security. One distributive option to explore is the use of minigrids.

6. System Grants. Grants should be made to alternative energy zones which are creating the energy infrastructure of the future. System grants should be aimed at researching how to best integrate the potential components of an alternative energy infrastructure through its interfaces and to improve the performance of the overall system.

So, if we can achieve the desired results from the interventions, the goal is feasible, but difficult. I might add that it's worth striving to achieve the desired results even if the effort falls short of the goal. Setting a realistic goal and aiming systematically for a feasible set of interrelated results will increase the chances of ending up with something close to what you want.

Identify Model Correction and Improvement Mechanisms

The baseline projections, the interventions proposed, and the projected results suggest what it would take to reduce imports by half in 20 years and reduce emissions in the U.S. to 40% of what they are today. They constitute a model of what could happen, based on common sense and on very rough, uncomplicated estimates. The model's projections are not a perfect representation of what will happen. No model is entirely accurate. But some can be very useful. The function of models and of measurement systems is not to compel rigid adherence to a plan, but to serve as a guide to action and a measure of progress.

By using models that have predictive baselines and measurement feedback, government agencies will have a better grasp of what is working and what is not working. Better informed, they can intervene earlier and more effectively, altering their emphasis and refining their approach, all to coax the market into producing what we need most at prices we can afford. This type of intervention uses models and measurement feedback to further a clear, simple, but profound guiding concept.

To increase the potential of this model and process to function as intended, some things need to be added. Strategists need to take into account potential problems in implementing their solutions. They need to look for improvement opportunities, anticipate risks, and make contingency plans. In addition, the model and selected measurements need to be continuously reviewed for effectiveness and adjusted to make them more effective; that is, to say the model needs a corrective mechanism. Correction mechanisms depend on measurement feedback that compares actual results to projected ones

An analogy can illustrate what is involved in improving a model through measurement feedback and correction. Suppose you decide to go on a diet. You recognize that weight loss comes from eating fewer calories than you use, After a month and a half of experimentation you find that if you eat 2500 calories a day, you lose 1 pound every two weeks. Using this history, you predict that you will lose 5 pounds every 10 weeks if you stick to the diet. You could create a little table to project what you expect to lose each week and then track your actual weight loss against the table. If you create such a table, you will have created a small model with desired results and a mechanism for monitoring and measuring your progress.

Let's say that this diet works quite well at first but that you begin to notice a change after a few months. You aren't losing as much weight as before. If you have been consistent in counting calories, you have to consider another explanation. A likely possibility is that you are reaching equilibrium between the calories you are taking in and the calories you are burning. If you want to lose more weight, you may have to either decrease calorie input or increase calorie output or both. If you haven't been exercising at all, you might adjust your weight loss model by adding two or three of hours of exercise a week and by checking what the effect is.

This type of reasoning is behind the model, baseline, measurement feedback, and correction I'm talking about. You check your predictions against factual feedback which represents reality, determine where you are off, analyze why, analyze the impact of corrections, make necessary corrections based on what you hypothesize is the cause of the problem, and check the corrections against new measurement feedback. Models should necessarily include a self-correction mechanism such as this.

Finding Alternative Routes to the Goal

The strategy I have proposed here may raise objections, especially with respect to the significant role given to coal in the short term. While I recognize the concern about coal pollution, I believe coal has to be considered because it is our most abundant hydrocarbon and fits within our existing infrastructure. The additional pollution from coal should be compensated for by

- The use of coal gasification and carbon dioxide sequestration
- The use of hybrid and electric cars
- Greater home energy efficiency
- Overall reduction in oil consumption

In fact, because coal gasification is more efficient than coal combustion and its emissions are significantly lower,[43] states should consider not only creating coal-to-oil conversion plants but replacing coal-fired power plants with coal gasification power plants to help reduce pollution levels. One study that supports this idea is FutureGen, a DOE project.[44]

However, I do worry that we might forestall using alternatives and continue to give coal the predominant role in the stationary power sector. We can prevent a lingering dependency on coal by careful planning that balances immediate needs against longer term objectives. We may, as an alternative, discard coal gasification as a solution and try to shorten the time it takes to ramp up production of fuel efficient cars, especially plug-in hybrid and electric cars. But one question would be, can our electrical grid accommodate these cars and can car manufacturers transition to fuel efficient cars more quickly than I've suggested?

Finding technological solutions to complicated environmental, political, and social problems requires a clear understanding of the problems we're trying to solve, openness to finding workable solutions, an ability to balance options, and the capability to find optimal paths to a solution. Can we achieve the goal of eliminating half our imports without coal? Can we achieve our moral environmental commitments with its use? How do we get the best return on the effort and money we spend? These are questions that more sophisticated models could help answer.

Whatever we do, we should be furthering the integration of the stationary power and transportation sectors through electricity together with the development of cleaner electrical power. This is the indispensable guiding concept we should refer to when looking for any near-term or longer-term solutions. It binds all the energy problems and addresses them together. Other concepts that just address climate change, or ecological damage, or energy shortages, or oil dependency separately can decouple the underlying problems and leave us with only partial solutions. We might be led, for example, just to stress coal-to-oil conversion, new oil extraction techniques, and nuclear power to avoid energy shortages without regard to pollution, or we could stress the reduction of pollution without addressing oil dependency.

Identify the Main Risk Drivers

The main risk drivers are likely to be sudden cataclysmic events and accelerated population growth. To guard against a sudden, huge drop in imported oil requires that every state have a contingency plan. I think these plans should have as their aim the ability to handle a 25% drop in oil imports, which today would be difficult to deal with.

To reduce accelerated population growth in the U.S. will take something that is possibly even more difficult, a change in immigration policies. Population growth is destabilizing many countries, and the unrest it causes is spreading to more stable nations, including those in the wealthy West. This dangerous natural acceleration may be difficult to recognize on a year-by-year basis, but consider this: Starting when humans first walked on the earth, it took until 1850 to reach 1 billion people. To reach 2 billion people took a little less than 100 years more (late 1940s). To reach 4 billion people took just 25 more years (1975).[45] What similar future increases will mean is staggering and life-changing. The likelihood of open world conflict over oil will increase greatly with increases in population. But people will be impacted in other smaller ways too. As an example, if individuals continue to drive alone to work in Los Angeles 20 years from now, one projection suggests it could take highways with 15 to 20 lanes to accommodate them. Continued immigration (legal and otherwise) at current levels is going to greatly exacerbate our problems.

Understandably, people from distressed countries want to emigrate to Europe and America for a better life. But they bring with them both an immediate increase in population and the threat of a higher birth rate, which will dash efforts in Europe and America to reduce birth rates and bring consumption of all sorts under control.

More broadly, any effort to aid distressed people, if it does not include an effort to control birth rates, will add to the instability, misery, and violence across the world. Unfortunately, countries, religions, businesses, and individuals resist birth control measures for a broad spectrum of reasons.

Corporations want increasing numbers of customers so they can grow. Continual growth and continually increasing consumption are the basis of our economic models today, which will have to change. Our planet can not sustain this growth. To keep prices competitive, corporations also want to pay the lowest wages they can. The more workers that compete for the same job the lower the wages are likely to be.

Countries want to be big and formidable. Religious groups want to increase the number of their adherents. Parents in third world countries want to have large numbers of children to ensure that they are taken care of in their old age. Some resist birth control, considering it a form of genocide even though the world population is in the billions and their populations are hardly going to disappear from birth control.

You cannot solve a problem just with wishful thinking or good intentions. You have to define the problem and its ramifications accurately to have a chance of finding solutions, and then you need to look for solutions that solve the problem, not some other problem. The alternative to accepting the reality of what we face and preparing for that reality is to lurch from emergency to emergency when the crisis comes, trying to improvise, trying to meet the demands, reasonable or not, of a panicking population, with every expectation that we will squander enormous amounts of money and get little in return for years. There shouldn't be any doubt that the costs of entering such a crisis state will be far greater than the costs of preventing it.

Looking Forward

This chapter has described what could be done to cut our dependency on imported oil by half in 20 years and reduce emissions by 40%. If we succeed, we will break the growing monopoly on the world's most vital commodity, oil, and free ourselves from the trap we've entered. The next chapter will describe how we can go farther.

27

The Alternative Energy Zone

Beyond the near-term campaign, we need a strategy for evolving our energy infrastructure so that we completely regain our energy independence while becoming more in harmony with nature. We should seek to end our dependency on foreign oil by mid-century. Alternative energy zones, which use the guiding concept of integrating the stationary power and transportation sectors through electricity, are a basis for such a strategy.

You'll recall that an alternative energy zone is a small community in which customers use alternative energy for their stationary power and transportation needs. For an historic and profoundly successful example of a zone, consider the first electrical power system that Nikola Tesla and George Westinghouse set up between Niagara Falls and Buffalo, New York over one hundred years ago. Tesla's inventive genius and Westinghouse's company made Buffalo a huge real-world laboratory for Niagara Falls' AC electricity and then extended the experiment to New York City.[1] Almost all the elements of the centralized grid system used throughout the country today were implemented in ten years in Buffalo and New York City, and the region between the two cities. The Buffalo experiment demonstrates how exposing a new energy system to real world conditions can accelerate its development and improvement. It is the evolutionary paradigm we should use to rapidly evolve an integrated alternative energy infrastructure from the existing stationary power infrastructure.

A strategy of implementing these zones has a number of benefits which can be seen from the growth of the Buffalo "zone." A zone is scalable. It allows participating companies, universities, and research labs to work within a relatively small but expandable geographical area. Because it is scalable, a zone is easier to manage, control, and evolve than a spread out system like a hydrogen highway. From the start, a zone engages a nongovernmental customer base that includes homeowners, car owners, and small businesses. With this customer base, companies can test the waters and gain real-world experience. Because companies will have a relatively large customer base to draw on, they can reduce the costs of alternative energy for customers in the zones and eventually for those outside the zones. As a further incentive, a zone will give alternative energy customers increasing energy independence. As it matures, alternative energy zones will prove that alternative energy works, demonstrate its possibilities to the wider community, and act as models for extending alternative energy across the country.

This chapter will describe how we could get started using alternative energy zones to evolve our current infrastructure into a new, integrated energy infrastructure. Since a very small minority will be in the alternative energy zones to start with, I'll also describe how interested individuals outside the zones can gain greater energy independence using some of the activities required to build a zone. But I'll start with activities exclusively related to building an alternative energy zone.

How to Build an Alternative Energy Zone

The first step is for some entity to decide it wants to initiate a zone. A power company (Westinghouse) created the initial AC power infrastructure in the U.S. But alternative energy zones don't have to be started by a company, they could be started by a university, a national laboratory, or a core group of customers themselves.[2] Good places to start the first alternative energy zones are in cities like San Francisco that have large numbers of pioneering "green" customers eager to try alternatives.

The core group should prepare an initial system model of the functioning zone, obtain financial and political support from local, state, and federal government officials, and build the zone using the model. An analogy is building a home. A home builder doesn't just start building. The builder starts by drawing a blueprint of the house to determine what will be needed to build it, then gets the money needed to meet the demands of the blueprint, and builds the house according to plan and city code.

A basic system model of an alternative energy zone is shown in Figure 36. Like the process diagram (Figure 34) in the previous chapter, it demonstrates conceptually how we might implement the guiding concept of sector integration. It is different than the previous models in that it is a depiction of what components might comprise an energy system and how these components might interact. It also shows an overlap (denoted by dashed lines in the figure) of the energy customers and the stationary power sector resulting from customers supplying electricity to the stationary power sector. The diagram is a high level view but for purposes of illustration it will provide a simple conceptual guide to building up a zone.

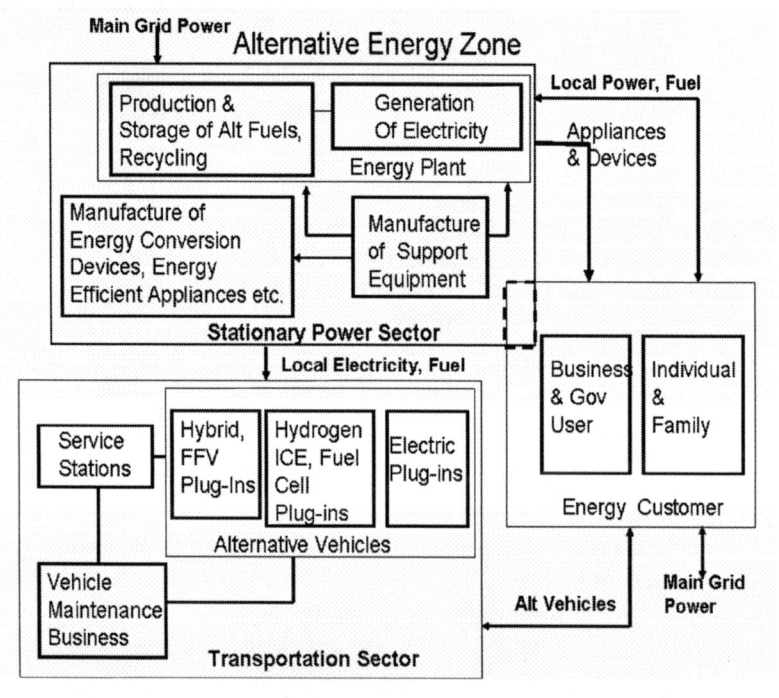

Figure 36. A Model of an Alternative Energy Zone[3]

The Role of Design in Alternative Energy Zone Development

There are a number of benefits to modeling a system beforehand, the main one being that you can determine what to offer and when to offer it. As a simple example, based on the model in Figure 36, the administrators of the zone would make hybrid plug-in cars and home solar PV systems available for purchase immediately. But they would make hydrogen ICE and fuel cell cars available only after the local production of hydrogen (and other alternative fuels) can be ensured. An energy plant that produces hydrogen would need to be established first, as one corner stone of the alternative fueling system. The energy plant would start out as a producer of power then could branch out to supply cars with hydrogen, ethanol, and other biofuels. The point is that knowing in advance what the components of a system are and how they might interact can help determine what sequence of implementation is most logical and economical.

By designing an alternative energy zone before it is set up, zone administrators can also require powerful product design heuristics that will help ensure competition and improvement. Two design heuristics to consider are modularization and loose coupling. Although these two concepts are drawn from computer science, they are readily applicable to alternative energy zones.

Modularization is a method of design which breaks up a system into components that can potentially be switched in and out readily, something like Lego building blocks. Modularization makes it easier to use standardized interchangeable parts. More than that, modularization can allow a product to grow in size and change or even be replaced, something which is much harder for huge monolithic systems to do.

Loose coupling is a design heuristic that allows designers to change some basic operational characteristics within a module without affecting the interaction between modules. Using loose coupling also isolates malfunctions and makes troubleshooting easier. To work, loose coupling requires modules to interact through a common interface. Input and output specifications for this common interface, though specific, have to be broad enough to minimize constraints on the operations within a module. Loose coupling with broadly specified interfaces eliminates much of the constraint that forces new products to be slavishly compatible with older ones or with one another within the wider system. It is easier to replace a product from one vendor with that of another, which helps to foster competition. Loose coupling, like modularization, can be applied across the zone from the highest conceptual level down to its smallest components.

Using modularization and loose coupling make it much easier to grow an energy system. Requiring the modularizing of solar panels using loose coupling, for example, could allow users to more easily expand their solar energy system or to change its components.

Guidelines for Building an Alternative Energy Zone

As the model begins to take shape, the core group (those who decide to initiate a zone) should establish how it should grow and evolve, how it should operate, who should lead the effort, and how the zone can obtain customers, among a number of considerations. Here are some sample guidelines to consider:

1. Choose Zone Administrators through an Electoral Process

As soon as the number of potential customers reaches a minimum number to implement a zone, use the electoral process to select the leaders, which consist of a zone director and staff members. They will be the zone administrators and should have specified roles, time of service, and compensation for overseeing the development of the zone.

2. Set up a Pre-Existing Customer Base and Boundary

The alternative energy zone should not be implemented until a pre-existing customer base of sufficient size has been established. The guideline for the model zone developed here is to begin with at least 150 customers. By customer I mean a business or a family or a group of tenants dwelling in the same building. Decisions have to be made also about the length of commitment with rewards for longer commitments.

Like the Buffalo city experiment, the boundaries of the alternative energy zone should also be defined. It might encompass a whole city or only a section of a city. As an example, consider a zone that would start in a section of a hypothetical city of 78,500 people. In this city there are 20,154 households of which 17,096 are families. The median age of the adult population over the age of 18 is 37. The average household income is $52,000. Unemployment is 4%. Eight percent of the family households have a single parent in the home. All of the single parents work. In 85% of the households with two parents, both parents work. The average number of children still living at home is 2.5. Approximately 18% of the working population works outside the city limits. About 5% of those who work outside the city limits use the transit system, which is centered at the downtown bus station. Building and construction have leveled off for now. There are 912 businesses in the city, of which 90% are small businesses, that is, businesses with 18 employees or less. There are 35 schools ranging from elementary to high school, and a university is nearby but outside the city limits. The area of the city is 17.5 square miles and is roughly rectangular.

The alternative energy zone might start in a section of this city and then grow to eventually encompass the entire city. Selecting customers who live and work in relatively close proximity to one another is a good way of establishing the first definable boundary.

3. Determine local resources

Once it catches on, every state may be expected to develop several zones, each of which takes into account the special energy resources of the local area. The energy resources of this sample zone would include sun, wind, sewage and other biomass. Nuclear power is not considered. Energy drawn from waves or currents can not be considered because the city is not near a coast or large river. Geothermal energy is not considered since none is locally available. Biomass may be considered as an energy source for the production of ethanol and hydrogen. The biofuels would be drawn largely from a local multipurpose plant that recycles trash, processes waste, and generates power. For solar power, an average of 5.5 peak hours of sunlight per day is assumed. For wind power, an average wind speed of 14 mph is assumed, with significant seasonal wind variation.

4. Model and Measure for Continuous Improvement

To ensure that the systems approach which underlies the zones will work, universities and national labs should be invited by the zone administrators to get involved. The universities and national labs can help model and simulate the zone before it is started and help create new products which industry can further develop for use within the zones.

Once the system begins to function, the zone should be continuously monitored by the zone administrator and staff who oversee the whole system, by the customers who use it, by the companies who were invited to create its products, by local and state government officials who promote it, and by the universities, national labs, and independent researchers who assist in engineering the zone and improving it.

Monitoring should include the use of metrics and measures that determine what progress is being made in decreasing the use of oil, increasing energy efficiency, reducing pollution, and improving products. The measures and metrics should help determine what is working and what won't work, what breakthroughs are needed and when to introduce them. They should provide reliable estimates of the system costs while helping to drive those costs down. And metrics and measures should be used to let customers know whether their actions are making a difference and whether they are saving money. The customers should receive from the zone administrators a quarterly account of the economic and ecological impact of their actions, including information on emissions.

5. Establish a Mechanism for Facilitating Improvements

Besides tracking progress, the zone administrators must establish and utilize a mechanism that facilitates the implementation of meaningful improvements suggested by customers, vendors, modelers, and product researchers.

6. Localize Energy Production and Distribution

To implement power generation, the zone administrators should consider the creation of a localized, mini-grid based on new Department of Energy (DOE) thinking. This is how a minigrid is described by the DOE.

> "One application of distributed energy (DE) is minigrids, a set of generators and local-reduction technologies that supply the entire electricity demand of a localized group of customers. By avoiding the cost of transmitting electricity from a distant central-station power plant or transporting fuel from a distant supply source, a minigrid (sometimes called a "microgrid") can significantly improve the economics of meeting energy needs..."[4]

If an alternative energy zone decides to use a minigrid as a means of localizing the power grid, the DOE should pay for its installation. Having independent power grids in local communities can put us on a more secure footing. A local stationary power system is analogous to our federalist system which localizes political power to an extraordinary extent, reflecting the independent spirit of our founding citizens. If the federal government were crippled, each state would be able to run itself independently. If both the state and federal government were crippled, each county and each city could continue to run for quite a while on its own. A similar benefit could be realized for an electrical infrastructure that is comprised of many local infrastructures that can operate independently when necessary.

Some other potential advantages of distributed power systems not mentioned by the DOE in the quote are:

- **Resilience.** Damage from disasters will be more simply contained. And those not directly affected by the damage will be in a better position to aid the recovery of those who are stricken.

- **Jobs.** Creating community-based, localized power grids and home-based power systems stimulates more community-based jobs.

- **Flexibility.** Individual and community power for home and business use can be scaled up over time while remaining localized.

- **Self-Sufficiency.** The conventional power grid can become more of a backup system than a primary system for the majority of alternative energy economic zones.

A major question will be affordability.

7. Leverage the Existing Infrastructure to Evolve a New One

The alternative energy zone should leverage the existing infrastructure even as it evolves into a more localized one. To be able to utilize the grid system and yet be independent of it when necessary, will require new smart grid technology. This technology should not only allow a community to run independently on its own local grid, it should also allow homes that have their own power systems to isolate themselves and continue operating during blackouts.

8. Put Energy Efficiency First

Everyone within the zone should expect to become more energy efficient before anything else is done. They should be using more energy efficient appliances and electrical devices, insulating their homes more effectively, and using passive solar, solar water heating, and solar thermal systems, if available, for heating and cooling. Within the zones, computers should be set up to monitor home energy expenditure and potentially to regulate it automatically. Zone members can also help to create practical, easy-to-use products and may suggest gadgets that will further energy efficiency. Being in a real-world laboratory can stimulate creativity in everyone who takes part!

9. Foster Solar Energy as a Major Source of Power

Whenever practical, solar energy should be considered as the predominant choice to produce electricity. Solar energy is our most abundant clean energy resource. It will foster the rise of the solar city, a symbol of an emerging alternative energy economy.

10. Prepare in Advance for Expansion

As the zone grows and matures, it will become more complex. The zone administrators should anticipate and prepare for expansion.

11. Encourage Competition to Ensure Quality and Reduce Cost

The alternative energy zone should include built-in competition. As the number of customers grows and the alternative energy zone expands and takes shape, so should the number of companies that compete for business. Companies should be judged on their ability to reduce costs and improve their products over time. Competitors that demonstrate that they can do a better job for less money should supplant their rivals as they would in any competitive marketplace.

12. Provide Judicious Subsidization

In this model, the system administrator should take action to reduce the price of alternatives using negotiated bulk pricing and then government subsidies as needed. The customers should be subsidized instead of the companies, so that real costs and prices can be easily tracked by zone monitors. A seven year payback rule could be used to ensure customers that they recoup most of their upfront costs. Based on payback, subsidies to further defray upfront costs could be determined. In return for price subsidies, customers will be expected to become product evaluators and to suggest improvements. For a reasonable additional maintenance fee and an agreement to continue evaluating system improvements on a regular basis, customers who have made valuable suggestions should be offered periodic system upgrades for the first five to seven years at least. Providing customers with updated versions of energy products they have helped improve should encourage customer commitment and should benefit both customers and manufacturers. For cars, similar rules should apply. In return for evaluation and improvement suggestions, the customer should receive an alternative vehicle that costs them no more than they will have spent on a comparable conventional car over seven years of use (including initial cost). Replacement vehicles should also be offered at a subsidized price to customers who have made suggestions that have led to improvements.

13. Plan for Diversification

In this model system, manufacturers of alternative energy products and product sellers, such as fueling stations, should be allowed to diversify the products they offer to increase their profit margin. Fueling stations may, for instance, continue to sell regular gasoline alongside hydrogen and blended fuels like E10 and E85. Diversification is another way of leveraging the existing infrastructure. Comparing alternative and conventional products in terms of price, number of sales, and customer satisfaction should be helpful in measuring the success of alternatives.

14. Ensure Improvement and Standardization through Association

Even though the initial alternative energy zones are each likely to start out with different models, members of different alternative energy zones are likely to find commonality if they communicate periodically. Administrators from different zones should review what they have in common, what they've learned works best, and start to standardize how they operate. Standardization can accelerate improvements. Newer zones will be able to tailor their own operations from common standards to fit their specific circumstances and can provide feedback to further accelerate improvements. And zones should consider creating a common website. A common website can, among other things, become a vehicle for increasing the power of zones to buy in bulk to reduce prices.

15. Anticipate Legal Issues and Provide Backup Systems

Alternative energy zones have a broad purpose. They are meant to create an energy micro-economy in which a diversity of potential alternative solutions are not only tried out but are interconnected and tested using a systems approach. Because mistakes will be made and risk is present, the zone administrators in this model should establish an adequate legal basis for the zone with state assistance, insulate the zone from unreasonable legal action, and provide adequate backup and safety plans.

To minimize risk and reduce inconvenience, stringent safety codes should be developed and enforced, and back up systems should be put in place early on. For instance, customers with experimental cars should have available conventional rental cars as spares free of charge – except for fuel costs – if a reasonable need arises, such as an occasional long trip in a hydrogen car where hydrogen isn't available or a malfunction of their alternative car.

Implementing a Model System

With these guidelines, the model in Figure 36, and the design suggestions in place, it's time to describe an illustrative implementation of an alternative energy zone, starting in the stationary power sector. Not all aspects of the implementation will be covered. The emphasis will be on what the zone administrators, customers, and interested individuals outside a zone can do to improve energy self-reliance.

The Stationary Power Sector

Before the zone administrators consider improving home and business energy efficiency and introducing local alternative power, they should focus on establishing a baseline of energy usage. Past energy bills show what customers have been paying and how much energy they have been consuming. By comparing past bills with new ones, customers will be able to determine the effectiveness of their efforts to become more energy efficient. Outside a zone, interested individuals may have to establish a baseline for energy usage on their own. They can get some help from the web and from regional utility companies that provide aid in determining a home's energy efficiency. To give a true picture, the energy bills need to cover a long enough period of time, at least a year, preferably several years.

Efficiency Starts at Home

A baseline energy survey of a home or business is the basis for an energy efficiency plan. The information collected from the monthly electric bill, the monthly natural gas bill, and any other obvious household and transportation energy bills is analyzed and the results of the analysis lead to recommendations. Typically, an energy surveyor will recommend that a set of energy inefficient appliances and electronic devices be replaced or improved to make them more energy efficient. The recommendations should be tailored to fit the customer's economic circumstances and needs, consider projections of potential growth, include a timeframe for action, and indicate a set of costs with any available subsidies figured in. The survey, if done within a zone, should also estimate reductions that can be achieved in the environmental pollution emitted by the home or business. Subsequent energy bills can be tracked by the zone administrators and compared to these estimates.

A good starting point is the examination of electric bills. As an example, in 2006 the consumption of electricity for an average 2,100 to 2,500 sq. ft. single family home was between 920 and 1540 kilowatt-hours per month or about 30 to 50 kilowatt-hours per day. How does this compare with someone's actual home electric bill? Table 6 has the actual costs for a 2100 square foot California home. This electric bill indicates that the sample household already consumes significantly less than the average amount of electricity.

There can be a number of reasons for this. One is the location of the home. Climate has an obvious influence on the amount of home energy consumed per year. Another consideration is the types of major appliances in the home. A homeowner may have reduced consumption of electricity by using natural gas for key appliances that could be run using either electricity or natural gas In the example case, the owners use natural gas to run the range, central heater, hot water heater, clothes washer, and clothes dryer. As a rule of thumb, the cost of running a heat producing appliance using natural gas is about half the cost of using electricity, and the use of natural gas for this purpose is more efficient.

Table 6. Baseline Power Consumption[5]

Month	Kilowatt Hours	Cost
Jan	411	$50.90
Feb	366	$46.58
Mar	357	$43.94
Apr	316	$37.30
May	437	$52.79
Jun	451	$53.74
Jul	579	$69.72
Aug	953	$126.90
Sep	771	$100.26
Oct	420	$50.42
Nov	382	$46.34
Dec	264	$32.43
Average	476	$62.83
Minimum	316	$37.30
Maximum	953	$126.90
Yearly Total	5707	$711.32

While examining the energy bills, you should determine what specific appliances should be replaced or modified to best increase home efficiency and reduce pollution. As an obvious common sense rule, you should be looking for the best energy payoff for the money.

The energy bills will help you make the determination of what appliances to target, but like a detective you have to do some sleuthing. An examination of the sample electric bill in Table 6, for example, reveals a seasonal pike in consumption due to the use of the air conditioner.[6]

Given that air conditioning increases energy usage significantly during the summer months, should an old air conditioner be replaced with the most efficient current model? Not necessarily. Replacing an older cooling system with a more efficient one is not the only option to make the home or business more efficient. There are other methods of improving efficiency such as installing a programmable thermostat. The point here is that replacement is not always the best option. There are relatively low cost, simple changes that can make a significant difference.

Although air conditioning is easy to spot as a big seasonal consumer of electricity, significant power consumption by other home appliances is more difficult to determine. Many utility companies now help their customers analyze their energy consumption but individual home owners and zone surveyors can check electrical consumption for themselves using a device called the "Kill a Watt." It determines exactly how much electricity an appliance or electronic device is consuming and can be purchased for about $50.[7] The results can be checked against the electric bill to determine the percentage of electricity each appliance is consuming.

Anyone can get a fairly reasonable estimate more cheaply using their energy bill and an online calculator set up for the purpose or an online guide such as *Appliance Energy Costs: costs for major household appliances*.[8] The guide reports the typical energy consumption of major appliances. It even provides some qualifiers to help consumers select examples close to what they actually have. For example, under the rubric *electric range*, the *Appliance Energy Costs* guide gives these choices: "Small surface unit, large surface unit, oven bake unit, broil unit, self-cleaning." The energy consumption naturally differs depending on the choices. Using a guide such as *Appliance Energy Costs*, a consumer can compare the totals from his or her monthly bills with the energy consumption estimates for each major appliance and electronic device. An even simpler alternative is to use government statistics.

Here is what I came up with working from the electric bill in Table 6, Energy Star's estimates of energy efficiency, and government statistics from Figure 35. First of all, the home has a 15 year old refrigerator. According to government statistics, adjusted just to consider electricity, the aging refrigerator uses approximately 13% of the total electricity consumed by the home (5% of the total energy from natural gas plus electricity). The other chief candidates to considered in this home are the 15 year old dishwasher, the five year old tower computers and their monitors, the 2 year old TV, VCR, and DVD, and lighting (including both outdoor and indoor lighting).[9] Table 7 suggests what the maximum annual energy savings might be.

Table 7: Potential Savings

Items	Percentage of Total Energy Consumption	Estimated Kilowatt hours Used	Max Efficiency Improvement	Max Energy Savings kWh
Air Conditioner (cooling)	18%	1024	50%	512
Dish Washer	5%	293	50%	146
Refrigerator	13%	732	50%	366
Electronics: including Computer with Monitor, TV/VCR/DVD	18%	1024	60%	615
Lights	26%	1463	75%	1098
Other*	21%	1171		
Totals	100%	5707		2736
		Overall Max Efficiency Improvement:		48%

* "Other" represents an array of household products, including stoves, ovens, microwaves, and small appliances like coffee makers and dehumidifiers.

The next step in a home energy survey is to put together the plan of action for the household. For example, as Table 7 suggests the energy surveyor might recommend that the family replace their aging 15 year old refrigerator with an Energy Star model that consumes less than fifty percent as much energy. The family might replace the out-of-date tower computers with laptop computers or mini-Macs, which are at least as capable and consume up to sixty percent less energy.[10] The family could replace the older and bulkier cathode ray tube (CRT) monitors with LCD monitors, which consume less than half the energy. For lighting, the family could replace their incandescent light bulbs with compact fluorescent lights (CFL) that use 75% less energy. The main point is to pay attention to the energy consumption of key appliances, which most people don't do, and to systematically replace them with newer versions that consume significantly less energy. Based on government statistics, replacement of all the items listed can reduce energy consumption by up to 48%.

Because heating and cooling have such a profound effect on energy consumption, both need to be considered in any home energy survey. Besides determining baseline heating and cooling consumption, everyone including those in this household can take some obvious actions to reduce consumption without diminishing comfort. If the household owners don't have a programmable thermostat, for example, they should get one. Increasing insulation and changing some habits can also have a big impact. This family already sets their programmable thermostat to 78°F during the day in summer and 70°F during the day in winter with similarly thought out adjustments during the night. Others might consider similar action. For hot water as well as for heating and cooling the house, the family should investigate the use of solar water heating systems or if available solar thermal systems that might provide both heat and electricity.

By making several of these changes, including some changes of habit, the average household can expect to reduce its overall energy bill significantly. The same is true for business owners who use energy surveys and assessments appropriate to their businesses to determine what they can do to save energy.

Within zones, it is especially important to have computer monitoring of a home or business energy system starting with a baseline. The zone should also explore the potential for computer control of the entire home system to improve efficiency further.

Generating Power On-Site

After the energy assessment is completed and the energy efficiency plan is largely implemented, the zone administrators and customers in the alternative energy zone (Figure 36), and interested customers outside the zone, would take the next step. They would select an alternative energy power system. If a minigrid is planned, they may elect to use it exclusively and not consider home power systems. Or they may consider using the minigrid as a backup to a home power system. Let's assume most will want an on-site grid-tied generator. One of the advantages of making the home more efficient is that a home energy system can be smaller and cost less.

Customers who want to own their own home energy system would select an alternative energy vendor whose system can provide the household's electricity needs. The electric bill can establish what constitutes 100% of the energy required, but figuring that out can be tricky. Just as an example, let's assume that after replacing inefficient appliances and electronic devices with more efficient ones, improving insulation, installing a programmable thermostat, and replacing their hot water heater with an Energy Star version, a family finds it consumes an average of 450 kWh per month. Acquiring an energy system that produces 450 kWh per month can work so long as the alternative system is grid-tied because the grid supplies additional electricity when needed and takes any excess electricity when it is produced. However, if anyone wants off-grid energy independence, average consumption is not a good measure. The heaviest monthly energy consumption and the capacity to store excess electricity for use later have to be taken into account and some margin added. Other considerations may include a future increase in consumption such as the addition of a plug-in hybrid or electric car.

In this example, home energy systems will be used and will be tied to the minigrid when it becomes available. The minigrid provides a cost effective backup to improve the overall capability of alternative energy to deliver electricity when needed during times of peak consumption. The minigrid will in turn have the main grid as its backup.

Both the minigrid and the home energy systems will have the capability to unplug themselves from their backup power source and run independently once the zone matures. But before the minigrid is established and before a home energy system can be automatically insulated from blackouts, customers who use a grid-tied home energy system will have to rely on the main grid. This means that they aren't allowed to continue generating electricity during a grid blackout due to possible adverse effects to grid workers and the grid from any electricity that they may inadvertently supply to the grid during a blackout.

As part of their duties, the zone administrators will be responsible for reviewing options with the customers and helping them choose what best fits their needs and budget. They will also need to consult with the zone customers regarding potential vendors ahead of time and gain some consensus in order to utilize their bulk purchasing power. The zone director also needs to ensure that home and small business owners receive money back for contributing excess electricity to the power grid or minigrid. This arrangement, called net metering,[11] is crucial. Net metering constitutes one of the means by which customers will reduce their overall energy costs and get back their upfront costs.

Home and Business Based Energy Systems

What could you expect if you purchased a solar energy or wind system for your home or business? The basic items of an on-site clean energy system are:[12]

THE ALTERNATIVE ENERGY ZONE

- Solar panels or a wind turbine
- Installation hardware

The first decision that a customer inside or outside the zone will have to make is whether to use solar panels or wind turbines or something else. The main consideration is usually cost and energy availability within a region. Is there enough wind, sunlight, river/ocean current or geothermal energy? Zone administrators will have studied the general situation for the zone and be able to advise customers.

To determine the available amount of sunlight, zone administrators will consult a sun map. The sun map in Figure 37 from the National Renewable Energy Laboratory (NREL) is an example of what is used.

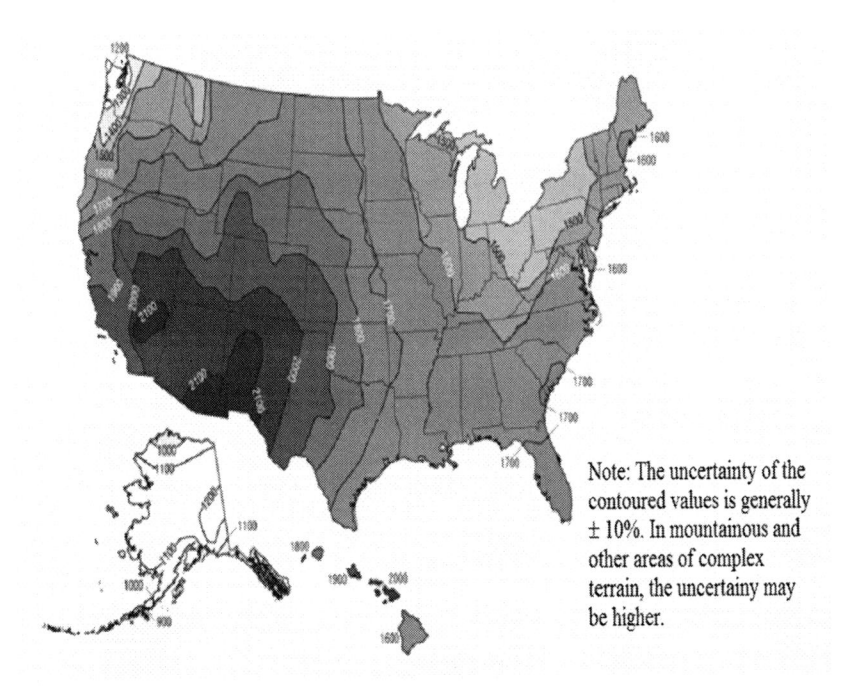

Note: The uncertainty of the contoured values is generally ± 10%. In mountainous and other areas of complex terrain, the uncertainy may be higher.

Figure 37. Sun Map[13]

Solar maps provide vital information about the kilowatt generating power of sunlight in various parts of the country. There is around 5.5 peak hours a day. The darker the area on the map, the more sunlight a region gets annually. The small numbers on the contours of these maps indicate the average rate of incident sunlight energy per unit area in terms of watt-hours per square meter per day.

Solar panel vendors like Southwest Power, Inc. and British Petroleum (BP) usually have their own sun maps and accompanying tables that approximate how much energy is available for use and what size system you'll need. As an example, assume you have a 2,500 square foot single-family home and use around 1240 kWh per month. Then Southwest Power's GT9000 is the only one of their systems that can possibly deliver 100% of the energy needs, and then only if the home is located in the Southwest. But if you increase the energy efficiency of your home by 30 to 50%, you'll naturally have more options.

You'll have more options too, if you use alternative energy for less than 100% of your needs and the grid for the rest. Today many grid-tied solar energy users derive around 70% of their power from solar or wind energy. If solar energy conversion systems are modularized to allow for easy expansion, energy generation capacity can be added over time, which may make solar energy more affordable. It can be a good strategy for both vendors and their customers.

A way you can figure out for yourself what size system you need is to use your electric bill to determine your average monthly energy consumption requirements and use an online sun map[14] to determine how much sun exposure you get. Then use the rule-of-thumb that a solar panel can generate 70 milliwatts per square inch (mW/in^2). As an example, assume a family after improvements consumes an average of 15 kWh per day and that they get an average peak sun exposure of 5.5 hours a day. Then this family, if they want 100% average consumption coverage would need at least a 270 square foot solar panel.[15]

If you select commercially available solar panels, you'll have some other decisions to make, such as where to put the solar panels. If you don't put the solar panels on the roof, you can choose between adjustable and non-adjustable mounting racks.

Solar panels on mounting racks that are raised up on poles can often double as shelter for cars or equipment and can provide electricity directly to a parked hybrid plug-in car through an outlet. The conversion efficiency of home solar panels without an adjustable mounting rack is 11 to 18%. Many systems today get around 15%. Adjustable systems can increase the conversion efficiency significantly; the amount varies. A more efficient adjustable solar panel reduces the size of the system

THE ALTERNATIVE ENERGY ZONE   293

required. But adjustable systems can also increase price because of the additional parts required.

Adjustable mounting racks can be either automatic or manual. An automatic system adjusts the mounting rack throughout the day. A manual system, which is cheaper, is adjusted seasonally.

If you decide to put the solar panels on the roof, you are likely to be provided with a non-adjustable mounting rack. But you still may have an option. Instead of a non-adjustable mounting rack, another option available today is solar panels that are a material part of the roof. Solar energy systems that are embedded as part of the roofing are currently limited to flat top and shingle roofs, but that will change.

While solar energy may be the energy source of choice for individual residents and cities throughout much of the U.S., in some northern portions of the country, as the solar energy map indicates, solar energy may not be a good choice. Wind may be the preferred choice. The wind map in Figure 38, from the NREL indicates the electric power potential of wind in various regions of the country.

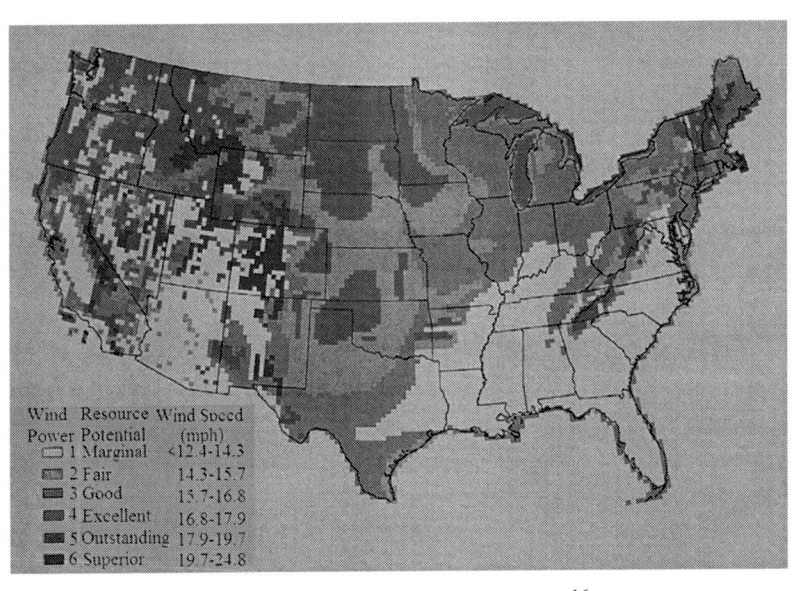

Figure 38. Wind Resource Map[16]

Fortunately, most regions of the country that don't have enough sunlight to generate the electricity they need tend to have adequate wind.

You can get a rough estimate of the wind in your area even without a wind map simply by looking for the following characteristics.

Estimating Wind Speed by Wind Effect[17]

MPH	Characteristics
0-1	Smoke rises vertically.
2-3	Smoke drift indicates the direction of the wind.
4-7	You can feel the wind in your face. You see leaves rustle.
8-13	Leaves and twigs are in constant motion; small branches move.
14-18	Small trees sway and white caps can be seen on ocean waves

Most commercially available wind turbines become productive at wind speeds of 14 mph and up. If your observations suggest that you are in a promising area, you might consider taking the next step and check local weather websites for information. Wind data can be found for most states, down to areas of a few square miles in some cases.

The main consideration is usually the wind speed averaged over a year. Basically, if your area has an average wind speed of 14 mph or more, a wind turbine may be considered as a source of power.[18] You may still consider a wind turbine if you have dependable 14 mph winds seasonally over roughly half the year.

In areas where a wind system can work, wind speed is typically one of three major determiners of how much power a wind turbine can generate. The other two are the wind tower height and the wind turbine's rotor size.[19]

There are also non-technical considerations, such as permits, that can affect the decision to have a wind system. In residential areas the rotor size[20] and tower height may be restricted, or wind turbines of any kind may not even be allowed. It is one of the jobs of an alternative energy zone administrator to find out this information and negotiate with city officials regarding permits. Outside a zone, individuals will need to investigate the possibilities for themselves.

A few manufacturers produce small wind turbines that can fit urban constraints and city ordinances, and more are on the way. An example is the Whisper 200 (H-80), which has a 10-foot rotor with an 80 square foot sweep area. Despite its small size, it can generate about 250 kWh of electricity per month for a wind speed of 14 mph[21] and around 320 kWh per month[22] for a wind speed of 16 mph. [23] An output of 250 to 320 kWh per month from a single wind turbine is below the level required to provide the electrical power needs for the average 2,100 sq. ft. house. But if you consume 920 kWh per month to start with, and reduce your energy consumption by one third, then two wind turbines would cover at least 80% of the household's needs, with conventional energy sources providing whatever else is needed.[24]

Alternatively, in many regions, a hybrid solar-wind energy system might be utilized to cover 100% of average monthly electrical consumption.

Whatever the needs and energy source – whether sun, wind, wave, current, geothermal, biomass, or other – you will also require additional components to make the electricity that is generated available to a household or business. Those components are:

- DC disconnect (PV array disconnect)
- Kilowatt-hour (kWh) meter
- Monitoring software
- AC panel distributor (switch box or breaker panel)
- Inverter
- Transfer Switch

The **DC disconnect** is used to safely interrupt the flow of electricity from solar panels or wind turbines, usually so you can add to or upgrade a system or troubleshoot it. When solar panels are used, the disconnect may be called a **PV array disconnect**.

A **kilowatt-hour meter** allows the utility company to keep track of electricity transferred from the grid to the house or business, and excess electricity transferred from the house to the grid. The utility companies in many states already provide these meters at no cost.

Monitoring software is used to determine how much energy is being produced by the on-site generator and consumed by the home or business. For a zone, it can provide important information about the zone's generation and consumption of electricity. It can also assist in regulating electricity between the grid and the on-site generator, and in improving energy efficiency.

The **AC panel distributor**, which is also called a **breaker panel** or **switching box**, already exists in every home and building that uses AC electricity. It is usually a wall-mounted box that is installed in a utility room, basement, garage, or on an exterior wall of the house or building. Inside the AC distributor are a number of labeled circuit breakers through which electricity received from the grid or an on-site generator is routed to the various rooms of a home or business. The breakers allow electricity to be disconnected for servicing and also protect the building's wiring against electrical fires by isolating any electrical overload. Utility companies may require an additional AC disconnect between inverters and the grid. The AC disconnect is located near the utility kWh meter.

An **inverter** converts the DC electricity produced by solar panels or wind turbines or some other type of DC power generator into the AC electricity commonly used in most homes and businesses. Home energy systems need a grid-tied inverter if the home or business is connected to the utility company's grid. The grid-tied inverter synchronizes the AC electricity converted from the DC generated on-site, with the grid's AC electricity. This synchronization allows the on-site system to feed excess electricity to the grid through the AC breaker panel. Over the last few years, inverters have improved significantly. They have made it possible for on-site systems to produce commercial grade electricity.

A **transfer switch** may be used to switch between home generated power to grid power or vice versa or it may be incorporated in an inverter. It can be automatic or manual.

Energy Storage Devices and Backup Generators

If a home or business is to be independent of the grid, it will require an energy storage system or other form of backup system that can seamlessly provide electricity when needed. If an energy storage system is going to be used, several components will need to be added to the solar panel, wind turbine, or hybrid system:

- Battery bank or other energy storage device
- Energy storage disconnect (battery disconnect)
- Charge controller
- Backup generator

Battery banks are a main choice offered today for use with solar and wind systems. Sealed absorbent glass mat (AGM) batteries are among the most desirable because they are maintenance free and designed for both grid-tied and grid-independent systems, but other types may be considered.

The **energy storage disconnect**, which today is usually a **battery disconnect,** makes it possible to safely isolate the energy storage device from the rest of an electrical system for servicing, modification, additions, and troubleshooting.

The **charge controller** protects a battery bank or other related energy storage system from overcharging. Secondarily, most charge controllers employ maximum power point tracking (MPPT), which is a tracking system that optimizes solar panel and wind turbine output. The charge controller performs its primary function by monitoring the energy storage device and interrupting the flow of electricity to it when it is fully charged.

With a grid-tied system, the excess electricity produced on-site after the energy storage device is fully charged would be sent to the grid. For an off-grid system, excess generated electricity could be lost if the battery bank isn't large enough so it's important to have a large enough battery bank for independent systems.

If there are times when there is not enough electricity stored or generated by the solar or wind system, a **backup generator** may be required to supplement these systems. A hookup would be needed to connect the backup system to the home electrical system.

Hybrid Systems with a Fuel Cell Backup Generator

For anyone interested in trying ultimate systems, one example would be a hybrid wind-solar generator with battery storage and a backup fuel cell generator. When the batteries are low and there is insufficient wind and sunlight, the auxiliary fuel cell backup generator could be used.

To create such a system, additional components would have to be added to a solar panel, wind turbine, inverter, AC distributor, battery bank, battery disconnect, and kWh meter. Those additional components are:

- Controller
- Fuel cell backup generator (or fuel cell car)

Figure 39 illustrates how all these components might be combined.

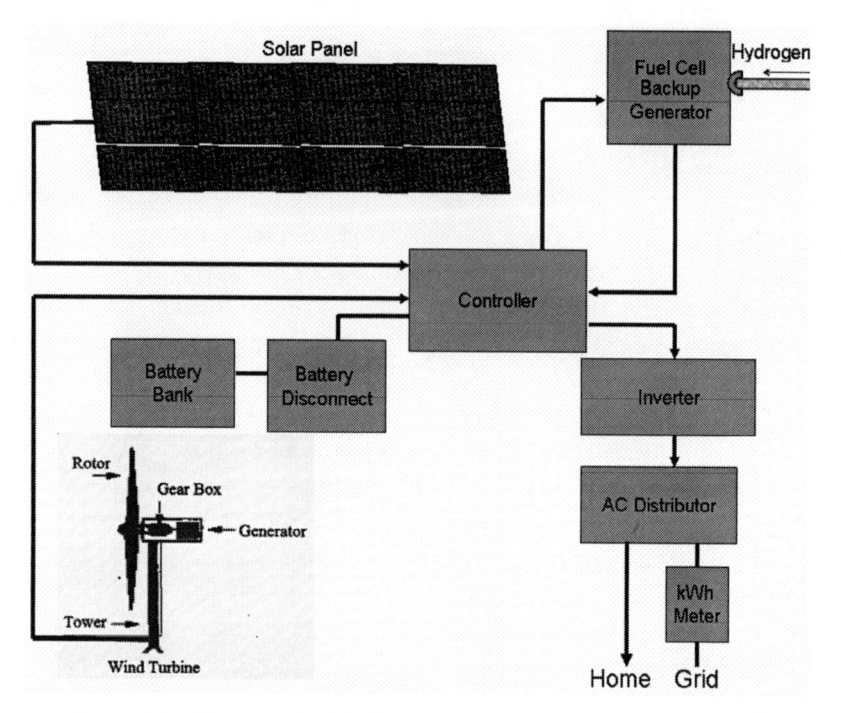

Figure 39. Home Wind/Solar Hybrid System with Fuel Cell[25]

Not shown but important is a computer that monitors the energy system.

How would such a system work? The hub of this system is a **controller** which would include charge controllers for the solar panel and wind turbine, a DC disconnect, and various other components for routing electricity. Inputs to the controller would be DC electricity from the solar panels, the wind turbine, the batteries, and the backup generator. Output from the controller would be DC electricity routed to the inverter or the batteries. Usually DC electricity would be routed to the inverter for home use. Excess DC electricity would be routed to the batteries or to the grid through the inverter. As an alternative backup, the electricity generating capability of a modified plug-in hybrid car might be used, although the plug-in car would normally receive electricity from the home system.

This configuration would be among the most advanced systems within a zone and would be a model for integration of localized stationary power with transportation, but there are other possibilities in the future.

Hybrid Systems with a Flywheel

Flywheels store kinetic energy within a rapidly spinning rotor or disk. They have been used in various forms for centuries and have a long history of use in automotive applications as speed moderators.[26]

A more advanced use for flywheels is storing energy that can be later used for generating electricity. When a flywheel's rotation is increased, energy is invested in it. A relatively friction free environment can be created by suspending the flywheel in a magnetic field. The energy that is stored in the rapidly rotating flywheel can be retrieved either mechanically or electromagnetically. In present day experimental prototypes, a motor/generator is mounted on the flywheel's shaft to either increase its rotational speed or to convert its rotational energy to electrical energy. These actions are analogous to charging and discharging a battery. A high-strength containment structure houses the rapidly rotating elements and low-energy-loss bearings stabilize the shaft.

Flywheels show promise as environmentally friendly energy storage devices, but there are some drawbacks. So far, their energy density is low. Also, there may be a possibility of an explosion if energy is released too quickly, so a home flywheel system needs to be placed well away from the house. Still they are already an option. There are flywheel systems commercially available. One example is made by a company called RPM which can be found on the web. RPM has a diagram showing the integration of its flywheel system with a hybrid solar/wind home energy system.[27]

Like flywheels and fuel cell backup generators, super capacitors and heat storage systems are possible technological innovations that could also be used to back up a primary alternative energy system, and more innovations are coming. The zones are the place to find out what will work and to improve capabilities.

Home Power Magazine

More generally, if you don't like existing commercially available choices or prefer to do-it-yourself, one place where you might look for information is *Home Power Magazine*. The publishers will allow you to download a free issue of their online magazine. It's worth reading an issue or two even if you aren't a do-it-yourselfer.

If you are a potential do-it-yourselfer and outside a zone, you may consider establishing a green club to pool experience and talent. At a minimum, you can learn a lot. Optimally, who knows. Many important companies have been founded in garages by groups of individuals who got together to experiment.

More on Continuous Product Improvement

In my approach, electricity is the key to the entire alternative energy system, but economics is the bottom line. Already, full reliance on home or community-based energy systems is within reach. But the breakout of this system to the wider community will depend on breakthroughs and improvements in quality achieved at a lower cost. There are many avenues to reducing costs and increasing efficiency. All can be explored from within alternative energy zones.

Building Alternative Energy into the Transportation Sector

In my model system (Figure 36) two to three basic types of alternative cars would be made available to customers to build up transportation as part of an integrated stationary power and transportation system. They have been described functionally in *Part Two: America at the Crossroads*. They are:

1. A gasoline electric hybrid plug-in car that is flexible fuel ready
2. An all electric car that doesn't use fuel cells
3. An electric car that does use fuel cells or
   a hybrid electric car with a hydrogen internal combustion engine

To qualify initially, each car must get the equivalent of at least 45 miles per gallon, and emit at least 80% less pollution than a conventional gasoline car. Hydrogen or biofuel city buses could participate in the alternative energy zone assuming they meet initial qualifications.

Some of the alternative cars will come with a solar "blanket" or an equivalent that allows their batteries to be partially recharged when exposed to light. Car manufacturers may wish to experiment with built-in solar panels on cars or an equivalent rather than with a solar blanket. An advantage of a built-in system is that the car can be charged whenever it is exposed to light whether the car is stationary or moving.

How much would recharging with a solar blanket or its equivalent help? The University of Queensland has created an ultra-lightweight, low drag, hybrid car that uses a 375 W solar blanket to recharge its 360V Li-ion battery pack.[28] They claim that their car can travel up to 37 miles a day on solar power alone. However, an ordinary car is heavier and consequently can't get anything like 37 miles a day from solar blanket recharging. For example, a Toyota RAV4-EV SUV with a solar blanket that absorbed sunlight in a parking lot for 5.5 hours could go about 7 miles on that charge.[29] Not far.

Still, a solar blanket may be worth considering. Once people begin to use a solar blanket, there will be more motivation to improve it. Furthermore, while the main driver is cost, there are advantages that go beyond individual cost. How much, for example, would the use of improved solar blankets reduce expenditures on the power grid?

While the transportation choices and their improvement are important, in the model zone I'm presenting, a major focus will be on building up viable, energy efficient alternative fuel production, storage, and distribution capabilities for the cars using biofuels and electricity. Production, storage, and distribution of alternative fuels are emphasized because they need the most improvement.

Real-world experimentation should quickly weed out unrealistic ideas about production, storage, and distribution, and lead to more practical solutions. For example, I doubt that piping hydrogen gas hundreds of miles from a hydrogen manufacturing plant to fueling stations is going to work. It certainly would be expensive. Yet some government estimators argue rather persuasively that it is actually the cheaper choice among a number of unpalatable choices.[30]

Furthermore, dispensing liquid or gaseous hydrogen through a nozzle and storing small amounts of hydrogen gas in large heavy fuel tanks on automobiles also seems wasteful and inefficient to me. Just because conventional cars are fueled from pumps, and liquid gasoline is used as fuel, doesn't mean the same paradigm should be used for hydrogen. I think there are better solutions which alternative energy zones could explore.

A Cartridge System

One possibility that could replace dispensing liquid or gaseous hydrogen is removable cartridges that contain a hydrogen absorbing material that would render the hydrogen virtually nonflammable. The cartridges would be designed for use in an easily accessible car storage compartment that would replace the conventional fuel tank of a car.

Is this feasible? Cartridges that hold hydrogen in lightweight hydrogen absorbing material already exist.[31] Mechanisms for forcing a hydrogen absorbing material to give up its hydrogen at a controlled rate are already well established. Cartridge fuel indicators are another matter. But even if cartridge fuel indicators can't be used, estimates based on use may be sufficient. Estimators would need the starting amount and a mechanism for correctly tracking usage.

A more basic challenge is storing enough hydrogen to allow a hybrid electric hydrogen car to travel without changing cartridges for a distance approaching what a conventional car can travel without refueling.[32] On-board storage of 6 kilograms of hydrogen would allow a driver to travel around 270 miles before running out of fuel. By comparison, a conventional car tank holds between 15 and 18 gallons of gasoline. If a conventional car's tank holds 15 gallons and its fuel efficiency is 21 mpg, it can go about 315 miles before running out of fuel.

How close are any manufacturers to having a solid-hydrogen storage system that can hold 6 kilograms? Texaco Ovonics Hydrogen Systems (TOHS) claimed in 2005 that the goal could be reached in 5 years.[33] Whether that is realistic or not remains to be seen. But the search for cheaper, more absorptive material for hydrogen has broadened significantly in the last few years, increasing the chances for a breakthrough. The National Renewable Energy Laboratory (NREL) is funding research into a wide variety of absorptive materials including complex hydrides, lithium amides, boron hydrides, aminoboranes, polyborane anions, nano particles, nanotubes, conducting polymers, carbon aerogels, glass microspheres, metal perihydrides, and clathrates. Both corporations and universities are involved in this effort.[34]

Let's assume this cartridge idea could be implemented. Why would anyone want it? A hydrogen containing cartridge has several advantages. Using the cartridges would make a hydrogen vehicle safer than a conventional car. Most of the hydrogen fuel would remain in the cartridge where it would be nonflammable. Only small amounts of flammable hydrogen would be used at a time in an engine or fuel cell. The cartridge could be standardized. Because hydrogen containing fuel cartridges would be nonflammable, they would be safe enough to be distributed by anyone through virtually any outlet and standardization of the cartridges would make possible mass production and distribution.[35] The cartridges could be bought at a local market, over the Internet, and at service stations; that is to say, the cartridge system could make use of the existing infrastructure.

Service stations would not have to change their pumping system to accommodate hydrogen since it wouldn't be dispensed by a pump. And the cartridge system would make it possible for car owners to install their own fuel.

Service Stations

Assuming that most people continued to use service stations, the process for using a hydrogen cartridge system would work this way. Participating fueling stations would keep a supply of pre-filled hydrogen cartridges on hand for customers. Because of the absorptive material in the cartridges, they would require less storage space than ordinary hydrogen gas tanks. The fueling station would need to keep a supply of hydrogen containing cartridges that is slightly larger than the peak number of hydrogen cars it services over a week's time plus the customers' depleted cartridges. The cartridge exchange should take no more than 3 minutes. That is only slightly more than the time it takes to fuel a conventional car.

If not near a fueling station, a customer could pull over to the side, take out the old cartridges, and swap in the spare pre-filled cartridges he keeps in his trunk with the spare tire. The depleted cartridges, as recyclables, would eventually be sent back to the local supplier for refill. For customers who wish to purchase supplies from a store or over the Internet, a week's supply at a time should suffice, and could be kept safely in the garage or a shed.

Fueling stations in the model alternative energy zone will be conventional service stations that continue to dispense gasoline to their conventional customers, but they will also provide hydrogen cartridges and include E10 and M10 and possibly an E85 and M85 gasoline-ethanol or gasoline-methanol or other biofuel mixes among the pump choices. The gasoline mixtures would be provided to the customer at the market price, but the customer would, as part of the payback plan, be subsequently given a rebate to ensure the price paid for the mixture is at or below the cost of gasoline. Government vehicles would be expected to use the same refueling procedures as the public. Public buses could make special arrangements with the local suppliers to keep cartridges on hand that are of their proper standardized size. Spare cartridges could be stowed on board the buses for emergency situations.

In the model (Figure 36), the number of fueling stations required to supply hydrogen cartridges would depend on the number of persons signing up to lease or purchase hydrogen vehicles.

Establishing Vehicle Maintenance Businesses

Another of the challenges for the model alternative energy zone (Figure 36) will be to set up effective alternative energy maintenance businesses and parts stores. The vendors of solar and wind equipment and the utility companies will be expected to become the chief maintainers and repairers of their systems. For cars, the zone administrator, city and state officials may need to enlist repair shops and service stations and help them to get special training in sufficient numbers to repair alternative cars.

Building an Energy Plant to Serve Customers

Another major problem is production of enough hydrogen and other alternative fuels. In the model (Figure 36), a local energy plant will be built in the stationary power sector to provide alternative fuels and to generate electricity. The challenge will be to make such a plant cost effective.

The new energy power plant, if it uses the modular design heuristic, can build in flexibility to control costs and enhance revenue; that is, it can add on capability over time on a pre-determined schedule that keeps pace with the growth of the alternative energy zone or surpasses the zone's growth; or the plant could cut back on construction if it needs to. This flexibility mitigates inaccuracies in early estimations of use. The power plant would be expected not only to serve the local alternative energy zone, but the wider local community.

Besides flexibility, multiple-use should be among the criteria the zone administrators establish prior to making a selection among prospective plant builders and operators; that is, a major aim should be to build a plant that can use solid waste and waste in water to produce hydrogen and other alternative fuels and electricity while cleaning up the environment. A recycling plant or capability should be installed on the

THE ALTERNATIVE ENERGY ZONE

plant's grounds. What would tie the multi-use capabilities together would be energy production.

Questions to consider in making a selection of a multi-use power plant might include:

Input: Will the plant use sewage and other recycled biodegradable waste products that are locally available?

Output: Can the proposed plant produce a diversified line of energy products such as hydrogen, syngas, methanol, ethanol, and electricity?

Usage: Will the plant be able to handle a diverse customer base?

Flexible, Modular Design; Cost Effective Expansion: Will the plant have a modular design? Can it expand cheaply and efficiently?

Cost: Will the projected cost of running the plant be competitive with the cost of running a conventional coal burning plant, given a multiple use strategy?

Initial Production Capacity: Will the plant meet the needs projected by the zone administrator and wider community and will there be some spare production margin?

Efficiency: Would the plant have at least a 3:1 EPR, outputs to inputs? Does the plant have a mechanism for increasing efficiency?

Environmental Impact: Would the plant have low emissions of toxic chemicals, at least 60% less than conventional equivalents, and at least 60% fewer green house gas emissions?

Transportability: Is the distance from the plant to the zone less than 30 miles?

Safety: Will the plant have a plan in place to satisfy potential safety concerns? The plan should ensure that the plant conforms to community safety standards.

Land Requirements: Is land for the plant available for purchase?

Maturity of Technology: Does the company have a track record? Does it have at least a prototype plant in operation that can provide realistic data for examination and certification and to correlate with projected costs of a new plant?

Universities and national labs could participate in making these determinations. Of course, the selection criteria need to be prioritized and weighted, given the likelihood that no one company would be better than its rivals in meeting all criteria.

A question that may arise is, why include biofuels like ethanol? I'd say their use in a real world laboratory, such as an alternative energy zone, could stimulate more rapid breakthroughs in production capacity and in EPR, which allows us to include them in a diversified fuel strategy. Furthermore, a car that can run on 10% ethanol can run with fewer harmful additives[36] thanks to ethanol's oxygenation of gasoline, engine cleaning, and knock reduction, something that Midgley and others discovered in the 1920s.[37]

In addition to meeting the proposed criteria, the multiple purpose stationary power plant owners should satisfy the zone director, staff, university, and local government that all costs and any weaknesses and constraints have been disclosed.

If no one company can be found that qualifies, I would suggest looking at a combination of companies and technologies that taken together could transform a city's garbage, sewage, water, and industrial waste into sufficient quantities of syngas (synthetic natural gas), ethanol, hydrogen, and electricity to meet the model's needs at a reasonable cost. A major candidate to consider for inclusion in this plant is the BEAMR, the microbial fuel cell mentioned in *Part II America at the Crossroads*.

The hydrogen from the BEAMR could be used in the operation of a hybrid SOFC - gas turbine power plant and eventually for fuel cell cars. And the waste cleaning capabilities of the BEAMR could be used to purify water.

Large Solar Power Plants and Wind Farms

In regions of the country that can not produce enough of their own alternative energy, alternative energy zones may wish to consider working with nearby counties or states to meet their alternative energy stationary power needs. Large county or state power plants that use solar energy might make more sense for other reasons as well. If a large solar energy power plant were included, then I'd expect concentrating solar power (CSP) technologies to be major competitors, given their actual and projected capacity. Sandia Labs engineers have estimated that a CSP system which covers a desert area 10 miles by 15 miles could provide 20,000 megawatts of power. They have estimated a CSP system covering an area 100 miles by 100 miles could theoretically meet the electrical needs of the entire United States at the current time.[38] While such a huge centralized solar farm may not make sense, smaller sized CSP plants may make a lot of sense in some desert regions of the country.[39]

Another option is a county wind farm. They may be an excellent choice as a source of electricity initially for a local community and ultimately for a broader sector of a state or several states. In the Midwest, small wind farms are already becoming a fixture, typically on farmland. They provide the farmer with additional income without taking up much land area. In California wind turbines are big business. The state operates 13,000 wind turbines in its three largest farms: Altamont Pass east of San Francisco; Tehachapi, south of Bakersfield; and San Gorgonio near Palm Springs.[40] In 2004, wind energy in California provided 4,258 million kilowatt-hours of electricity, which is about 1.5 percent of the state's total. It is enough to light a city the size of San Francisco.[41]

The California Energy Commission estimates that newer wind technology will reduce the cost of wind energy to around 3.5 cents per kilowatt-hour.[42] That will make wind turbines a very economical choice for generating electricity. However, in the areas of California where the three largest wind farms were placed, wind speeds are generally highest in the spring and hot summer months. Approximately ¾ of their energy output occurs during those months. The seasonal variation of wind is one of the limitations of wind farms in most parts of the country.

The main limitation can be quantified by looking at the wind capacity factor. It is sometimes called the load factor and is the capacity of an energy source to generate electricity over a period of time. For example, solar panels, which receive 4 ½ –6 ½ peak hours of sunlight on average during a 24 hour day have the capacity to supply electricity around 19 % of the time. Wind turbines in areas that average 14 mile per hour winds will typically have a capacity factor of around 20% with 30% considered outstanding[43] Combining wind and solar energy farms can be a more effective alternative than either alone as it is for individual homes, but there may be cost issues.

There are a few places where the capacity factor is such that wind farms are likely to be the best answer to the entire state's energy needs. North and South Dakota are an example. The Dakotas have been called the Saudi Arabia of wind because of their extraordinary capacity factor. You can get some idea by looking back at the wind map, which shows wind conditions from good to outstanding. These conditions exist virtually all year long, which means the capacity factor for a wind farm is above 30%.

Extending Energy Efficiency

A number of suggestions have already been made for how we can use alternative energy to change our world. I would end with a humble final example: the separation of trash. The alternative energy zone should promote the separation of trash for recycling and the recycling of bigger items as a counter reaction to the extreme wastefulness of our society. It doesn't take much of a change in habit to separate out plastic, aluminum, and glass from other "waste" products. Recycling is less expensive than refining raw material from scratch because it takes less energy. Recycling conserves important metals such as nickel and copper, and it should result in less pollution.

In a number of states recycling programs already exist. To encourage recycling as well as other life-style changes, the community needs to be aware of the savings and the good they are doing.

Embedded in the concept of alternative energy zones is a philosophy of wise energy use. Relatively minor changes in life-style and habit can make a big difference. Energy conservation is both a good starting point and ending point for discussing how we can secure our future.

28

The Bottom Line

The bottom line from a customer's perspective and that of anyone seeking more energy independence today is cost. What will it cost? To a significant degree that will depend on the extent of participation by individuals, the communities, the state, and the federal government, in an alternative energy movement. The greater the collective participation, the less the individual and the government should expect to pay even with no further breakthroughs in technology. It is possible to get some idea of the outer boundary of costs by looking at a limited worse case scenario and situation.

Consider an individual family buying what they need at current prices. Assume they don't have the benefit of an alternative energy zone, a green club, or other organization to help them purchase their system at bulk prices with special discounts. Assume the members of this hypothetical family:

- Are not do-it-yourselfers.
- Have decided to replace at least some of their older appliances with more energy efficient ones. On their list for possible replacement are:

  – 15-year-old clothes dryer
  – 15-year old dishwasher
  – 15-year-old refrigerator
  – 10-year old T.V.
  – 5-year-old PC computer with 5-year-old monitor
  – 21 incandescent light bulbs

- Have decided to derive 100% of their average monthly electricity consumption from solar modules tied to the grid.

- Have a ten-year-old car, which they want to replace with a hybrid vehicle[1]
- Average 5 peak hours of sunlight in their area
- Will consume an average of 450 kWh per month of electricity after making their home more energy efficient.

The Cost of replacing inefficient appliances

Given these assumptions, it's possible to estimate their expenses. The costs will vary from place to place and over time, but should appear reasonable to anyone shopping for new appliances and electronics in 2008.[2]

Energy efficient clothes dryer: $1,350
Energy efficient dish washer: $1,500
Energy efficient refrigerator: $2,500
LCD TV set 26-31 inch screen: $1,700
Mac-Mini with mouse & keyboard: $950
Flat screen monitor 19 inch: $500
PC style widescreen laptop with mouse: $1,500
21 energy efficient lights: $50
Total: $10,050

For a limited budget, the family can consider only the new appliances and electronic devices that are likely to have the biggest impact on energy bills given cost:

Energy efficient refrigerator: $2,500
Mac-mini with mouse & keyboard: $950
Flat screen monitor 19": $500
PC style widescreen laptop with mouse: $1,500
21 energy efficient lights: $50
Total: $5,500

The Cost of investing in a Home Energy System

For a solar home system, I'll select just as an example, a PV system made by Shell, that provides 100% of the home's average electrical needs.

THE BOTTOM LINE

This PV system example includes:

256 sq. ft. of solar cells
Inverter and DC disconnect
Wiring, hookup, and permits
25 year warranty
Meter
Installation labor

From these costs I subtract estimated rebates, discounts, and tax deductions allowed by the Federal government and an example state, California.

Total Cost of Home Energy System, before rebates: **$27,280**

Rebates and Tax credits   **($9,000)**
    California Energy Commission (CEC) Solar Rebate
    for a 3,150 watt system that generates 14.8 kWh
    per day: $7,000
    Residential Federal Tax Credit
    for a 3,150 watt system: $2,000

Net Cost of Solar Powered Home Energy System: **$18,280**

Let's say that the family opts to buy a solar PV system at a cost to them of $18,280. What the family also needs to consider is the payback due to monthly energy savings and net metering. Unfortunately, a family that is paying $750 a year for a monthly average of 450 kilowatt-hours of electricity would pay more for solar energy than for conventional grid electricity even after 20 years of use if costs remained constant. Costs of electricity are, of course, likely to go up. But that is not much of an inducement now.

The current costs of a home power system are why we must find ways to lower the price. Bulk buying is one answer. Whether in an alternative energy zone or working with neighbors and friends, or by some other means, increasing your buying power by increasing the number of purchasers at one time can help reduce costs.

The Cost of Buying a Hybrid

So far, hybrid cars are more cost effective than PV solar systems. As a snapshot, for 2006 there were eight or nine Toyota Prius "packages" with a price tag ranging from around $21,500 to almost $30,000. The packages available depended on the region of the country. It was an unnecessarily complicated marketing system. Not all the packages were offered in all regions of the country.

The base sticker price of $21,500 was about $4,000 dollars more than the cheaper, conventional gas guzzling cars that might be selected. Given these numbers, a Prius owner who drives 15,000 miles a year and pays $3.00 per gallon could make up the difference in price in four years. After 12 years, a Prius owner should save in gasoline expenses the equivalent of 61% of the base cost of his vehicle, assuming the gas guzzler he might have bought instead gets 21 miles per gallon and the Prius gets at least twice the miles per gallon. Estimates, of course, will vary according to a number of factors including how often and how far the Prius is driven, how it is driven, and what the road conditions are.

Adding to the attractiveness of the Prius, the government offered rebates or tax credits that amounted to around $3,500 for early buyers in 2006. Unfortunately, because of high demand and short supply of Prius hybrids in 2006, a number of dealers were able to jack up their price two to five thousand dollars over the sticker price. Prius chat rooms were often the best places to find out which dealers were jacking up the price and which were not.

As a Prius owner, I've found so far that maintenance costs for the Prius are less than those of a conventional car, but that may change when we have to replace the battery. Nonetheless, there are few cars that can claim to save you money over time if you purchase them instead of a conventional gasoline car. The Prius can make that claim.

It is expensive to get a new car or new appliances and electronic devices. Few people can afford these things all at once. It will take time. However, most people will eventually buy new appliances, new electronic devices, new cars, and more. If you invest in alternatives, you invest in an improved environment and help us all regain our energy independence.

29

A Call to Action

I have spent much of the book discussing how we got into our current situation, why we need alternatives to hydrocarbons, what we have available as alternatives, and how government and industry can influence you and me to change our energy habits. But in the end, we the people have to decide that making this change is worthwhile. And having decided that a change in our energy infrastructure is necessary, which I hope you are convinced it is, we have to set about with a will to make the change.

To succeed we must believe that individual actions matter. We must believe that it is not only in the nation's interest, but also in our own personal self-interest to regain our energy independence as our country once gained its national independence. I urge you to act. There is so much that is possible. Consider making your next car a hybrid. Push to make hybrid plug-in vehicles available. Make recycling a way of life. Replace worn out old appliances with Energy-Star energy efficient ones. Join others in your community in demanding that we phase in more and more electricity produced from alternative sources such as solar and wind energy and from waste products. Push your state to fund research in areas where we need breakthroughs such as more efficient, lower costing solar cells; higher density, longer lasting batteries; and more durable fuel cells. Work with others to establish an alternative energy zone or energy co-op or green club so you can buy at bulk prices. Even if you have never joined any organization before, consider joining an alternative energy advocacy group. And push your state government to pass tax laws and provide subsidies that favor green energy. None of us should sit out this struggle for energy independence and a clear environment. None of us should want other countries to decide for us what our energy future will be. None of us should want to stand aside and let our land become increasingly polluted.

Deciding to make the world a better place and acting to make a difference are noble actions, worthy of a free people living in a Great Republic.

Major References

1. *Energy: The Next Fifty Years*, Organization of Economic Co-Operation and Development (OECD), 1999.

2. Davidson, C.I., *Clean Hands: Clair Patterson's Crusade Against Environmental Lead Contamination*, Nova Science Publishers, Inc., May, 1999

3. Huber, Peter W. et al., *The Bottomless Well*, Basic Books, Perseus Books Group, New York, 2005.

4. Feldman, David Lewis editor, *The Energy Crisis*, John Hopkins University Press, Baltimore and London 1996.

5. Gold, Thomas, *The Deep Hot Biosphere*, Springer-Verlag Inc., New York, 1999.

6. Leeb, Stephen, *The Oil Factor*, Warner Business Books, New York and Boston, 2004.

7. Nakicenovic, Nebojsa et al., editors, *Global Energy Perspectives*, Cambridge University Press, 1998.

8. Romm, Joseph J., *The Hype about Hydrogen*, Island Press, Washington, London, 2004.

9. Sperling, Daniel, and Cannon, James S., *The Hydrogen Energy Transition*, Elsevier Academic Press, New York, London, 2004

10. Daniel Yergin, *The Prize: The Epic Quest for Oil, Money, and Power*, Simon & Schuster; January 15, 1991

11. Gibson Consulting Oil Statistics @http://www.gravmag.com/oil.html

12. *Hydrogen Posture Plan An Integrated Research, Development, and Demonstration Plan*, U.S. Department Of Energy, February 2004.

13. Kunstler, James, "The Long Emergency," Rolling Stone, March 2005. See http://www.countercurrents.org/po-Kunstler280305.htm\

14. *National Hydrogen Energy Roadmap*, November 2002, U.S. Department of Energy

15. *A National Vision of America's Transition to a Hydrogen Economy - To 2030 and Beyond*, February 2002, U.S. Department of Energy

16. Texaco Oxonics Hydrogen Systems, LLC Ovonic *Solid Hydrogen Storage*. See http://www.ovonic.com for more details.

17. Zubrin, Robert et al. *Report on the Construction and Operation of a Mars Methanol in situ Propellant Production Unit* See http://www.pioneerastro.com/Projects/Mmispp/MMISP_Phae_1_Final_Report.doc

END NOTES[1]

Note: Some of the references in the footnotes are to web pages on the internet that may become unavailable. These web pages might still be found by using "The Wayback Machine" of the internet archive website at http://www.archive.org/web/web.php

PREFACE

[1] From Gibson Consulting Oil Statistics:
http://www.gravmag.com/oil.html

[2] Ayatollah Ali Khamenei [BBC, Friday, 5 April, 2002, 20:21 GMT 21:21 UK].

[3] *The Gathering Storm* by Winston Churchill for parallels between today and the 1930s.

1. THE AMAZING TWENTIETH CENTURY

[1] The internal combustion engine developed over a period of years culminating in Otto's invention.

[2] Englishman Sir Humphry Davy invented the first electric carbon arc lamp in 1801. The first electric incandescent lamps were invented in the 1870s by Joseph Swann of England and Thomas A. Edison. The American Thomas A. Edison invented the first commercially successful incandescent lamp around 1879. By 1880, Broadway in New York had lighting from direct current (DC). But it was rare in homes.

[3] The induction motor is still widely used in household appliances. Tesla demonstrated his original two phase induction motor in a lecture at Columbia University in 1888. He quickly developed a poly (three) phase power system.

[4] Marconi got the credit for inventing the radio, but there is credible evidence supporting Tesla's claim to be the true inventor. The theoretical idea of wireless transmission of energy, however, was not a new idea. It had been previously proposed by James Clerk Maxwell.

[5] The patents included polyphase AC motors, AC generator, transformer, transmission lines, and lighting.

[6] Westinghouse would soon regret the royalty contract.

[7] DC suffered more progressive loss of power over distance than AC.

[8] Because AC current switches direction several times a second, it only flows through the surface of the wire. This effect is known as "the skin effect."

[9] Parts of New York were already being lit with DC electricity. But the set up was cumbersome.

[10] Author Illustration – S. L. Klein

[11] Rotating copper coil past stationary magnets also works to induce electric current in the copper coils.

[12] Different phases of current are depicted as sine waves, undulating waves that in their form resemble ocean waves. They peak and trough at different times relative to one another.

[13] A watery analogy to an electrical surge might be a tidal wave.

[14] A 1000 ft lightning bolt, according to wikipedia, carries around one gigavolt. http://en.wikipedia.org/wiki/Lightning

[15] Author Illustration – S. L. Klein

[16] And later nuclear power.

[17] There is some dispute over Cugnot's birth year; some put it in 1723.

[18] http://www.usroots.org/~genranch/rails/ for some historic details

[19] Thomas Davenport and Scotsman Robert Davidson were the first to use non-rechargeable electric cells.

[20] Etienne Lenoir patented the first practical internal combustion gas engine in Paris 1860. It ran on coal gas. But it really didn't work very well. It was slow and lumbering, and the engine timing wasn't very good. The gas was not compressed but drawn into the car's cylinder by the movement of the piston and ignited halfway through the stroke. Otto's invention is dated around 1876.

[21] http://www.worldwar1.com/heritage/marnetaxis.htm for more details. Europe was busy building its own hydrocarbon infrastructure just as the U.S. was. But Europe lagged behind in mass production.

END NOTES

[22] In the early Twentieth Century global warming, environmental pollution, and air pollution didn't seemed like compelling problems. There were a few who worried about eventual oil depletion, but the period in which the energy infrastructure was build saw multiple oil field discoveries, especially in the U.S.

[23] Author illustration – S. L. Klein

[24] Author Illustration – S. L. Klein

[25] One liter is equal to approximately 1.06 quarts.

[26] 746 watts = 1 horsepower. The term horsepower was coined by inventor James Watt.

[27] It could also be defined as raising 33,000 pounds one foot in one minute.

[28] A 2006 Honda Accord has 166 horsepower; a Dodge Durango has 210 horsepower; and a Toyota RAV4 SUV has 268 horsepower.

[29] The adaptation of the oil industry to serve the automobile has a biological analogy – what's called homologous structures – these are structures whose similarity comes from one preexisting structure evolving into another. Darwin's observation that lungs probably evolved from swim bladders is a famous biological example.

2. PILLARS OF SOCIETY

[1] Gibson Consulting Oil Statistics. See Major References.

[2] http://www.tf.uni-kiel.de/matwis/amat/def_en/kap_5/advanced/t5_1_4.html The process was independently discovered in 1851 by William Kelly.

[3] There are now a number of types of steel, with melting point temperatures varying from about 2500 to 2900 degrees. http://www.firesleuth.com/table/metal.htm

[4] Notably, Bessemer had used Bleanavon iron ore, which has the rare characteristic of being phosphorous free, so he was initially unaware of some obstacles to the successful use of his process.

[5] Hall applied for the patent in 1886; it was granted in 1889.

[6] http://www.geocities.com/bioelectrochemistry/hall.html

[7] Bauxite ore contains a large amount of aluminum hydroxide (Al2O3·3H2O), along with other compounds.

[8] http://inventors.about.com/od/pstartinventions/a/plastics.htm

[9] See the movie *The Graduate*.

[10] A polymer simply defined is something made of many units – like a chain. Each link in the chain is a "mer" or monomer, a basic unit. To make a polymer, many links or "mers" are hooked, that is, polymerized together.

3. A Place at the Table

[1] http://www.fertilizer.org/ifa/statistics/indicators/pocket_summary.asp Look for the heading "Fertilizer Consumption."

[2] http://en.wikipedia.org/wiki/Thomas_Malthus for additional details.

[3] http://www.fertilizer.org/ifa/statistics

[4] http://www.fertilizer.org/ifa/statistics/indicators/

[5] DDT stands for dichlorodiphenyl trichloroethane; BHC stands for benzene hexachloride.

[6] http://ipm.ncsu.edu/safety/factsheets/pestuse.pdf

Knutson, R. D., C. R. Taylor, J. B. Penson, & E. G. Smith, 1990. *Economic impacts of reduced chemical use*, Knutson & Associates, College Station, TX

4. Dependency and Paradox
None.

5. Oil and the Third Wave of Totalitarianism

[1] It can be argued that there is an inevitable evolution in human society toward democracy. I'm not so sanguine. Democracy can survive and may prevail, but its survival is not assured. The ethnic differences in

END NOTES

many societies are clearly one of the major threats. They can also be a strength.

[2] Both Hitler and Stalin proclaimed that their movements were historically inevitable.

[3] An exception is historian Daniel Yergin. See his *The Prize: The Epic Quest for Oil, Money, & Power*.

[4] The Communist demonstrators were joined by many non-Communists in their opposition to U.S. involvement in World War II. Since George Washington first warned Americans against entangling alliances, there has been an oscillation in American politics between isolationism and engagement. See Washington's Farewell Address of 1796.

[5] http://cta.ornl.gov/cta/Publications/Reports/ORNL_TM_2000_152.pdf for more details.

[6] In 2004 Abdul Qadeer Khan confessed to being involved in a clandestine international network that proliferated nuclear weapons technology to Libya, Iran, and North Korea

[7] The PBS series frontline had this to say about the madrassas: "Many of the Taliban were educated in Saudi-financed madrassas in Pakistan that teach Wahhabism, a particularly austere and rigid form of Islam which is rooted in Saudi Arabia."
http://www.pbs.org/wgbh/pages/frontline/shows/saudi/analyses/madrassas.html

[8] The overt reason for the massacre was the accusation that Armenians were aiding the Russian invaders during World War I. But attempts to exterminate them go back at least to 1894, according to some accounts.

[9] Robert Siegel's interview with John Lancaster of *The Washington Post* about the case of the Pakistani woman who was gang-raped as punishment for a crime for which her brother was convicted. Pakistan's Supreme Court under extreme international pressure overturned the acquittal of the 13 men directly involved and ordered their re-arrest for the rape of Mukhtaran Bibi.

http://www.npr.org/templates/story/story.php?storyId=4721974 also http://www.westernresistance.com/blog/archives/001885.html

[10] She committed suicide. After her death, Egypt became a Roman province.

[11] The Arabs had helped the Allies win their war with Germany and its ally Turkey. In return they expected independence not just from Turkey but from their Western allies.

[12] Edward Gibbon's date (476 A.D.) for the fall of Rome is usually accepted because in that year Odoacer deposed the last emperor to rule from Rome.

[13] This portrayal is not without irony given that Saladin was a Kurd. Saddam used poison gas among other lethal weapons to slaughter thousands of Kurds during his rule.

[14] Even though no embargo was imposed on France or Great Britain, both suffered from a general cut in production ordered by OPEC. Where an embargo was imposed, most notably on the Netherlands and Portugal, the hardships were even greater.

[15] Like the Armenian genocide.

[16] Jihadists are already very busy in Europe. They are not likely to wait, but they could win through the ballet. Several European countries including France have low birth rates outside the immigrant populations. If the current disparity in population growth continues, France is expected to have a Muslim majority by 2050, if not sooner. If the extreme Islamists gain control of even a small but sizable portion of this Muslim population, the clash between cultures could be very bloody.

[17] Both the Sunni Bin Laden and the Shiite Khomeni used fatwas like the mafia uses contract killing to eliminate opponents and to intimidate. Iran's former leader even posted a reward for the murder of a novelist whom he felt had offended the Koran.

[18] A letter expressing the intent of Al Qaeda's former Jordanian leader al-Zarqawi to spark a civil war in Iraq has been widely reported. See http://www.csmonitor.com/2004/0304/p03s01-woiq.html

[19] Besides instituting federalism, democracy could help bring peace through the rule of law which, if seen as even handed, could quell the blood feuds and tit-for-tat retaliation on both sides.

[20] The so-called "Oil for food" program administered by the U.N. is a good example. It is no wonder Saddam believed he could thumb his nose at the U.N. given how easily he could bribe the influential including the son of the man who then headed the U.N.

END NOTES

6. THE ENVIRONMENT, THE MAD HATTER, AND HIS DIRTY DISHES

[1] http://www.geo.ucsb.edu/~jeff/sb_69oilspill/69oilspill-articles2.html

[2] The oil fires lasted for months producing gagging black smoke and numerous hospital trips.

[3] In mines and caves, levels over 25,000 Bq/m^3 have been measured. Bq stands for becquerel, a unit of measure for radiation; m^3 stands for cubic meters.

[4] Author illustration – S. L. Klein

[5] Catalysts cause chemical reactions to occur without themselves being affected.

[6] The TRI was established by Section 313 of the Federal Emergency Planning and Community Right-To-Know Act of 1986 (EPCRA). Under EPCRA, manufacturing facilities in specific industries are required to report their environmental releases and chemical waste management annually to the U.S. EPA.

[7] http://www.scorecard.org/general/tri/tri_gen.html and http://www.epa.gov/tri/

[8] Researchers at the University of North Carolina School of Public Health have linked Benzene to many forms of leukemia including acute myelogenous leukemia AML acute and chronic lymphocytic and myeloid leukemia. (See Savitz, D., and Andrews, K., "Review of Epidemiologic Evidence on Benzene and Lymphatic and Hematopoietic Cancers," Amer. J. Industrial Health 31:287-295 (1997)).

[9] http://www.benzene-lawyer.com/html/conditions.html

[10] http://bss.sfsu.edu/raquelrp/projects/oil.ppt. This is a rough estimate. The average sized refinery is increasing: "…EIA's Petroleum Supply Annual, Volume 1 noted that although the number of refineries stayed the same between January 1, 2003 and January 1, 2004, capacity increased by 137,000 barrels per day, adding, again, the equivalent of another medium-sized refinery."

[11] http://www.ornl.gov/info/ornlreview/rev26-34/text/colmain.html

[12] Predictions are based on the following: Only trace quantities of uranium are found in coal, ranging from less than 1 part per million (ppm) to around 10 ppm. The amount of thorium contained in coal is

generally about 2.5 times greater. The normal concentration of fissionable uranium-235 is 0.71% of uranium content. The typical plant has an electrical output of 1000 megawatts. Existing coal-fired plants of this capacity annually burn about 4 million tons of coal each year. Using these data, and the projected population growth, both U.S. and worldwide fissionable uranium-235 and fertile nuclear material releases from coal combustion should increase dramatically over the next 40 years if nothing is done to transition from a fossil fuel economy to an alternative energy economy.

[13] http://www.ornl.gov/info/ornlreview/rev26-34/text/colmain.html

[14] According to the World Health Organization (WHO), motor vehicles account for about 30% of emissions of nitrogen oxides, 50% of hydrocarbons, 60% of lead, and 60% of carbon monoxide in cities of developed countries. In city centers the values rise to 95% for carbon monoxide and up to 70% for nitrogen oxides.

[15] http://www.epa.gov/air/airtrends/aqtrnd97/brochure/summ.html for details.

[16] Occupational Environment Med 2000:57:477-483 (July). http://oem.bmjjournals.com/cgi/full/57/7/477

[17] For example, $2NO \Rightarrow N_2 + O_2$ or $2NO_2 \Rightarrow N_2 + 2O_2$

[18] For example, $2CO + O_2 \Rightarrow 2CO_2$

[19] Author illustration – S. L. Klein

[20] http://chemcases.com/tel/tel-13.htm

[21] http://chemcases.com/tel/tel-13.htm

[22] Some insist there is still approximately 300 times more lead in the atmosphere than existed before 1923.

[23] http://www.epa.gov/oar/aqtrnd00/lead.html

[24] http://chemcases.com/tel/tel-13.htm

[25] "Trashing the Oceans" from *U. S. News and World Report* for some additional details.

http://www.mindfully.org/Plastic/Ocean/Trashing-Oceans

[26] http://www.ewg.org/reports/bodyburden1/ and http://www.advancedhealthplan.com/toxicbody.html

END NOTES   325

[27] Dioxins and furans have the same basic chemical structure, which includes chlorine atoms. There are 210 differentiated dioxins and furans of which a dioxin called **2**,3,7,8,-TCDD is considered the most toxic.

[28] These include DDT, chlordane, and related pesticides. They are man-made chlorinated, halogenated hydrocarbons that work mainly by blocking nerve impulses necessary for normal bodily functions. Unfortunately, they are long-lived, toxic to most animals, and convertible to even more deadly compounds as they degrade or are eaten and re-released into the environment as metabolites. DDE is an example of a metabolite of DDT.

[29] Phthalates have multiple uses. As plasticizers, they soften plastics; they are used as oily additives to perfumes and hairsprays. And they are used as lubricants and wood finishers. The new car smell, which becomes pungent after the car has sat in the sun for a few hours, is partly the odor of phthalates volatilizing from a hot plastic dashboard.

[30] http://www.chelationtherapyonline.com/technical/p111.htm
http://www.epa.gov/opptintr/library/pubs/archive/acct-archive/pubs/bench.pdf
http://www.ehponline.org/docs/2000/108-3/forum.html
http://www.crtk.org/library-newsletters-2.cfm?date=1995-11-01+00:00:00&docID=256

7. THE CLIMATE THRESHOLD

[1] http://www.nasa.gov/vision/earth/environment/2005: "2005 was the warmest year since the late 1800s, according to NASA scientists. 1998, 2002 and 2003 and 2004 followed as the next four warmest years."

[2] http://yosemite.epa.gov/OAR/globalwarming.nsf/content/Climate.html

[3] http://yosemite.epa.gov/oar/globalwarming.nsf/content/Climate.html for more details

[4] Often, estimates of greenhouse gas emissions are presented in units of millions of metric tons of carbon equivalents (MMTCE); each gas is weighted by its GWP value, or Global Warming Potential.

[5] http://yosemite.epa.gov/oar/globalwarming.nsf/content/climateuncertainties.htm

[6] http://usinfo.state.gov/gi/Archive/2006/Jan/26-309163.html

[7] We need to better understand the impact of what we have done, what we are doing, and what we plan to do. A major area of uncertainty is the behavior of clouds in the face of global warming. Will it moderate or accelerate climate change? The Jet Propulsion Lab (JPL) is among the institutions trying to find an answer. I cite this example simply to suggest we need as full an understanding as possible of the impact of what we are doing so we don't inadvertently accelerate global warming. Granted that we have to act without full knowledge - it wouldn't be prudent to wait. Nevertheless, we still need to learn as much as possible as quickly as possible.

[8] Higher temperatures decrease the concentration of oxygen in water.

[9] http://www.underwatertimes.com/news.php?article_id=07128691043 for more details.

[10] http://www.energybulletin.net/3647.html

[11] http://www.stanford.edu/dept/news/pr/97/970730schneider.html

8. POPULATION GROWTH BLOW OUT

[1] http://www.iiasa.ac.at/Research/LUC/Papers/gkh1/chap1.htm for more details.

[2] The tar sands of Alberta Canada will be discussed in the next chapter. Coal conversion will be discussed in Part II.

9. THE HIDDEN COSTS OF OIL

[1] http://eia.doe.gov/emeu/international/gas1.html

http://auto.howstuffworks.com/gas-price.htm/

[2] http://www.icta.org/press/release.cfm?news_id=12

[3] *"The Real Price Of Gasoline,"* CTA's categories were (1) Tax Subsidization of the Oil Industry; (2) Government Program Subsidies; (3) Protection Costs Involved in Oil Shipment and Motor Vehicle Services; (4) Environmental, Health, and Social Costs of Gasoline

END NOTES

Usage; and (5) Other Important Externalities of Motor Vehicle Use. See http://www.icta.org/doc/Real%20Price%20of%20Gasoline.pdf

[4] http://www.ndcf.org/, and http://www.iags.org/n1030034.htm *NDCF report: America's Achilles Heel the Hidden Costs of Imported Oil* in http://ndcf.homeip.net/ndcf/energy/NDCF_Hidden_Costs_of_Imported_ Oil.pdf.

[5] John L. Moore, Carl E. Behrens, John E. Blodgett , *Oil Imports: An Overview and Update of Economic and Security Effects*, December 12, 199 @ http://www.cnie.org/nle/crsreports/energy/eng-53.cfm

[6] John L. Moore, Carl E. Behrens, John E. Blodgett , *Oil Imports: An Overview and Update of Economic and Security Effects*, December 12, 1997. section "Oil Imports as a Policy Concern."

@ http://www.cnie.org/nle/crsreports/energy/eng-53.cfm

[7] http://www.montanaforum.com/rednews/2003/11/02/build/accountability/warcost.php?nnn=6

[8] John L. Moore, Carl E. Behrens, John E. Blodgett , *Oil Imports: An Overview and Update of Economic and Security Effects*, December 12, 1997. See "the Military Costs and Imported Oil." See the "Measuring Persian Gulf Security Costs."
@ http://www.cnie.org/nle/crsreports/energy/eng-53.cfm

[9] See *Oil Imports: An Overview and Update of Economic and Security Effects*, December 12, 1997

@ http://www.cnie.org/nle/crsreports/energy/eng-53.cfm

See "Measuring Persian Gulf Security Costs."

[10] *Oil Imports: An Overview and Update of Economic and Security Effects*, December 12, 1997, Appendix.

See http://www.cnie.org/nle/crsreports/energy/eng-53.cfm

[11] There are other individual studies I've seen – one clearly linking benzene with rare cancers and others linking particulate matter in diesel exhaust, for instance, to some lung diseases. But we need more visible, more comprehensive studies.

[12] *Oil Imports: An Overview and Update of Economic and Security Effects*, December 12, 1997. See "Environmental Costs and Imported Oil." @ http://www.cnie.org/nle/crsreports/energy/eng-53.cfm

[13] *Oil Imports: An Overview and Update of Economic and Security Effects*, December 12, 1997, see Appendix @ http://www.cnie.org/nle/crsreports/energy/eng-53.cfm

[14] *America's Achilles Heel the Hidden Costs of Imported Oil.* See "Cumulative Impact."

[15] http://www.iags.org/costofoil.html for more details.

[16] *America's Achilles Heel the Hidden Costs of Imported Oil.* See "Cumulative Impact."

[17] *Costs of Oil Dependence: A 2000 Update*, David L. Green, Nataliya I. Tishchishyna, prepared for the Oak Ridge National Laboratory May 2000, ORNL/TM-2000/152. See Introduction.

[18] *Costs of Oil Dependence: A 2000 Update*, David L. Green, Nataliya I. Tishchishyna, prepared for the Oak Ridge National Laboratory May 2000, ORNL/TM-2000/152. See "Conclusions."

[19] *America's Achilles Heel the Hidden Costs of Imported Oil.* See the section entitled "Economic Impacts." The estimate appears to have been made in 1987 dollars.

[20] This was estimated using the year 2000 as the benchmark for dollar value. Costs of Oil Dependence: A 2000 Update. See "Conclusions."

[21] http://www.gravmag.com/oil.html for more details.

[22] Daniel Yergin, *The Prize: the Epic Quest for Oil, Money, and Power*, Simon & Schuster; January 15, 1991.

[23] *Oil Imports: An Overview and Update of Economic and Security Effects*, December 12, 1997, see end of section "Economic Costs and Imported Oil," http://www.cnie.org/nle/crsreports/energy/eng-53.cfm

[24] *Oil Imports: An Overview and Update of Economic and Security Effects*, December 12, 1997, see section Economic Costs and Imported Oil., http://www.cnie.org/nle/crsreports/energy/eng-53.cfm

[25] John L. Moore, Carl E. Behrens, John E. Blodgett, *Oil Imports: An Overview and Update of Economic and Security Effects*, December 12, 1997 @http://www.cnie.org/nle/crsreports/energy/eng-53.cfm
See section on "Measuring Persian Gulf Security Costs."

[26] http://www.eia.doe.gov/emeu/cabs/Usa/Oil.html

END NOTES

Also see www.arb.ca.gov/msprog/zevprog/CleanEnergy/Moorer.pdf and http://www.worldoil.com/Magazine/MAGAZINE_DETAIL.asp?ART_ID=2593&MONTH_YEAR=May-2005

[27] Hellman, Karl H et al. *Light-Duty Automotive Technology and Fuel Economy Trends: 1975 Through 2004*, Advanced Technology Division Office of Transportation and Air Quality, U.S. Environmental Protection Agency EPA420-R-04-001, April 2004

[28] http://www.epa.gov/otaq/cert/mpg/fetrends/420s05001.pdf

[29] http://www.atsnn.com/story/41277.html

[30] http://www.atsnn.com/story/41277.html
http://www.eia.doe.gov/neic/press/press235.html
http://www.worldoil.com/Magazine/MAGAZINE_DETAIL.asp?ART_ID=2593&MONTH_YEAR=May-2005

[31] http://www.eia.doe.gov/neic/press/press235.html

[32] http://www.sierraclub.org/population/reports/water.asp
http://www.iags.org/la020204.htm
"The Sino-Saudi Connection: INDIA'S ENERGY SECURITY CHALLENGE"
http://www.iags.org/n0121043.htm,www.embavenezus.org/pag_energy_pdvsa_may5_2003.php
http://www.netl.doe.gov/publications/proceedings/96/96jpfs/jpfs_pdf/indcon.pdf

[33] Venezuela is currently challenging the assumption that the Middle East has the world's largest reserves, but to date has not substantiated its claim.

[34] http://www.iags.org/geopolitics.html

[35] http://www.bp.com/subsection.do?categoryId=95&contentId=2006480 for an example review.

[36] http://bp.com/sectiongenericcarticle.do?categoryid=9003054&contentid=7005895

[37] http://www.atsnn.com/story/41277.html

[38] http://www.iags.org/n1030034.htm

[39] http://www.iags.org/n1030034.htm

[40] http://www.atsnn.com/story/41277.html

[41] http://www.atsnn.com/story/41277.html
http://www.eia.doe.gov/oiaf/ieo/pdf/world.pdf
http://www.energy.senate.gov/legislation/energybill/charts/chart8.pdf

[42] http://www.atsnn.com/story/41277.html

[43] Tar sands" or oil sands are a mud-like material composed principally of sand, water, and clay, wrapped in a thick hydrocarbon called bitumen. Bitumen is a type of low grade petroleum usually heavily interlaced with non-hydrocarbons such as nitrogen, sulfur, oxygen, nickel, and vanadium.

[44] http://www.worldenergy.org/wecgeis/publications/reports/ser/bitumen/bitumen.asp

[45] http://www.atsnn.com/story/41277.html

Also see *Can Canadian sands replace Arabia's?* @ http://www.energypulse.net/centers/article/article_display.cfm?a_id=421

[46] The interview at http://www.atsnn.com/story/41277.html.

[47] *Twilight in the Desert: the Coming Saudi Oil Shock and the World Economy*, Matthew R. Simmons, Wiley Publisher, June 2005.

[48] http://www.marcon.com/marcon2c.cfm?SectionGroupsID=51&PageID=464

[49] *Forecast of Rising Oil Demand Challenges Tired Saudi Fields* by Jeff Gerth (NYT).
http://query.mytimes.com/gst/abstract.htmlPres=FOOD10F93CS80C778EDDA80894DC404482
http://www.atsnn.com/story/41277.html

10. BACK TO THE FUTURE

[1] For more information on trophic cascades, see *The Song of the Dodo, Island Biogeography in an Age of Extinction* by David Quammen.

[2] http://english.aljazeera.net dated November 2, 2004

[3] http://www.vheadline.com/neadnews.asp?id=25948
dated February 28, 2005

END NOTES 331

[4] http://bbc.co.uk dated December 20, 2005

[5] BBC News http://news.bbc.com.uk/2/hi/middle_east/4749812.stm dated February 25, 2006.

[6] http://chinadaily.com.cn dated: May 15, 2006.

11. ALTERNATIVE TECHNOLOGIES

[1] "Powering the Planet" by Nate Lewis, *Engineering and Science*, Volume LXX, Number 2, 2007, California Institute of Technology, pg. 19

12. BUILDING BLOCKS AND STEPPING STONES IN TRANSPORTATION

[1] Author illustration – S. L. Klein

[2] Author Illustration – S. L. Klein

[3] For example, in May 2007 in Southern California, the average cost of gasoline was $3.25 per gallon. According to the EIA the cost of electricity in California was 14.3 cents per kWh. If a conventional gasoline car gets 21 miles per gallon then the cost to go 100 miles at $3.25 per gallon would be $15.48. In contrast, the cost of the electricity needed to fully recharge an electric car with a range of 100 miles on a single charge of 27.4 kWh, would be about $3.92. That is about 1/4 the cost of gasoline for the trip. (Calculation: 27.4 kWh x $0.143 / kWh = $3.92

[4] http://en.wikipedia.org/wiki/Toyota_RAV4_EV

[5] 248 horse power or approximately 186 kW. You can convert horse power to kilowatts by multiplying by 0.75.

[6] http://www.teslamotors.com/engineering/how_it_works.php.

[7] The first battery was created in 1800 by an Italian Alexandro Volta using an arrangement known as a voltaic pile. It consisted of layers of zinc, blotting paper, and silver in a jar. The electrolyte used was salt water.

[8] The battery will be ruined if deeply discharged. On the other hand, it benefits from partial discharge.

[9] http://www.eere.energy.gov/cleancities/hev/hev_components.html

[10] http://www.eere.energy.gov/cleancities/hev/hev_components.html

[11] UCLA reported is working on an "air hybrid car," yet another design variation. http://www.engineer.ucla.edu/sotries/2004/hybrid.htm

[12] Author Illustration – S. L. Klein

[13] http://www.eere.energy.gov/afdc/vehicles/hybrid_electric_what_is.html

[14] http://www.fueleconomy.gov/feg/hybrid_sbs_cars.shtml

[15] Aerodynamic drag (D) is equal to ½ density of the air (d) times the frontal area of a moving shape (A) times the velocity of the object (V) squared or ½ dAV² For a car, D is in Pounds, V in mph; d =391 times the drag coefficient.

[16] The Prius, however, is a bit hefty. It weighs around 2,765 pounds (1,255 kg).

[17] Photo by author.

[18] Author Illustration - S. L. Klein

[19] The description is based on public information supplied by Toyota. See Toyota's *A Guide to Hybrid Synergy Drive*

[20] Over half a million Prius hybrids were sold worldwide in 2006.

[21] Also known as liquid petroleum gas

[22] http://www.fueleconomy.gov/feg/bifueltech.shtml

[23] http://www.liberalartsandcrafts.net/contentcatalog/autos/altfuels.shtml

[24] http://www.isuzu.co.jp/world/technology/randd/projects3/01.html for some details.

13. BIOFUELS AND COAL: CATALYSTS OF TRANSITON?

[1] EPR is also sometimes called the Energy Profit Ratio.

[2] The EPR calculation is a bit involved. The energy input, for example, takes into account energy expenditure starting from scratch.

In the case of producing ethanol from corn, it would consider the energy to produce fertilizer to grow the corn, tractors to harvest it, trucks to

END NOTES

transport it to plant, etc. For information on energy it takes to produce ethanol, http://www1.eere.energy.gov/biomass/net_energy_balance.html.

[3] *Ethanol Production Using Corn, Switchgrass, and Wood; Biodiesel Production Using Soybean and Sunflower*, by David Pimentel and Tad W Patzek, Natural Resources Research, Vol 14. No1, March 2005.

[4] http://www.poe-news.com/stories.php?poeurlid=48314

[5] http://www.sequiturs.com/global_issues/energy/energy_profit_ratio.html

[6] *Electricity generation from cysteine in a microbial fuel cell* by Bruce E. Logan et al, Water Research 39 (2005) 942-952, received 28 July, 2004.

[7] Booki Mina, Shaoan Chenga, Bruce E. Logan, "Electricity generation using membrane and salt bridge microbial fuel cells," Pennsylvania State University, 2 February 2005.

[8] http://www.sciencedaily.com/releases/2004/02/040224081342.htm

[9] http://www.washingtonpost.com/wpdyn/content/article/2006/08/19/AR2006081900842.html

Note however that ethanol is 30% less efficient than gasoline, something which should be figured into costs.

[10] http://www.cornandsoybeandigest.com/mag/

[11] http://cornandsoybeandigest.com/mag/

[12] http://www.futurepundit.com/archives/003032.html

[13] http://www.gasandoil.corn/goc/features/fex43159.htm. The chemical process: $(2N+1)H_2 + NCO \rightarrow C_n H_{2n+1}+N H_2O$. A similar process is discussed in the Chapter on hydrogen production.

[14] The LTC process was invented by Lewis C Karrick who worked for the U.S. Bureau of Mines.

See Robert A Nelson's article

http://www.rexresearch.com/karrick/karric-1.htm

14. BROADENING THE STATIONARY POWER BASE

[1] http://www.awea.org/faq/wwt_basics.html

[2] http://www.homepower.com/ for more information on wind power.

[3] http://www1.eere.energy.gov/windandhydro/wind_how.html

[4] dB = decibels – is the unit of measure for sound.

[5] http://www.sequiturs.com/global_issues/energy/energy_profit_ratio.html

[6] The semiconductor's real potential was not realized until the 20th Century.

[7] Current + Voltage = Wattage (also known as Power). This is really an approximation.

[8] Other types of concentrated solar cells are also being explored. A Fresnel lens is produced by a company called Pyronsolar. http://www.pyronsolar.com Spectrolab, a subsidiary of Boeing, is using multi-layer semiconductors that resemble those used by JPL. See http://www.Spectrolab.com. Spectrolab's results will be discussed in more detail later.

[9] Author illustration – S. L. Klein

[10] This Dish System is used as a 10 kW water pumping system. It was produced by WG Associates for use by Native Americans in the Southwest U.S. Given their expressed belief in being in harmony with nature, it should not be surprising to find that Native Americans are among the early adopters of clean energy technologies.
See www.eere.energy.gov/consumerinfo/pdfs/**solar**_overview.pdf

[11] http://www.energylan.sandia.gov/sunlab/PDFs/solar_dish.pdf

[12] http://www.energylan.sandia.gov/sunlab/PDFs/solar_dish.pdf

[13] Author Illustration – S. L. Klein

[14] http://www.eia.doe.gov/kids/energyfacts/sources/renewable/
http://www.powerfromthesun.net/chapter1/Chapter1.htm, and
http://www.eere.energy.gov/troughnet/pdfs/solar_overview.pdf

[15] http://www.energylan.sandia.gov/sunlab/PDFs/solar_trough.pdf and http://www.powerfromthesun.net/chapter1/Chapter1.htm and http://www.eere.energy.gov/troughnet/pdfs/solar_overview.pdf

[16] http://www.energylan.sandia.gov/sunlab/PDFs/overview.htm#dish

[17] http://www.wipp.ws/science/energy/powertower.htm

http://www.cbc.ca/stories/2002/08/21/aus_power_020821

END NOTES

[18] 40% Efficient metamorphic GainP/GainAs/Ge multijunction solar cells, by R.R. King et al., Applied Physics Letters, 90, 183516 (2007)

[19] http://www.eere.energy.gov/RE/ocean.html

[20] http://www.energyocean.com/press01142005.html
http://news.bbc.co.uk/1/hi/sci/tech/1032148.stm

[21] The quote is from Alan Wallace, the co-principal investigator at Oregon State University (OSU)

[22] http://www.mt.luth.se/hydropower/BEST99/others.html

[23] Author Illustration – S. L. Klein

[24] http://www.mt.luth.se/hydropower/BEST99/others.html

[25] The estimated number of homes that can be serviced depends on the estimate of how much electricity is consumed in the average home. Figures range from 920 to 1540kW per month for an average home in the U.S.
http://tonto.eia.doe.gov/ask/electricity_faqs.asp#electricity_use_home

[26] The oceans' mechanical energy comes from different sources than the ocean's thermal energy. Whereas the ocean heat comes primarily from the sun with contributions from geothermal vents, tides are driven primarily by the gravitational pull of the moon, and waves are driven primarily by the winds and sometimes by ocean currents.

[27] http://www.scjai.com/technote149.html

[28] siemens-foundation.org/en/competition/2004_winners/aaron_goldin.htm

[29] "Powering the Planet" by Nate Lewis, *Engineering and Science*, Volume LXX, Number 2, 2007, California Institute of Technology, p. 19

[30] National Geographic online
http://news.nationalgeographic.com/news/2005/01/0114_050114_solarplastic.html

[31] For more details on wave, tidal, and current energy, checkout the EPRI website http://www.epri.com/oceanenergy/oceanenergy.html#briefings and
http://www.epri.com/oceanenergy/attachments/ocean/briefing/June_22_OceanEnergy.pdf

15. A PARALLEL PATH TO THE FUTURE

[1] See *Hydrogen Posture Plan An Integrated Research, Development and Demonstration Plan*, February 2004, pg. 2.

16. THE HYDROGEN INTERNAL COMBUSTION ENGINE

[1] Isotopes are atoms of the same element that differ in the number of neutrons they have.

[2] The coldest temperature ever recorded on Earth was approximately −129 ° Fahrenheit.

[3] Check Figure 2. Internal Combustion Engine Cylinder to review components of the cylinder.

[4] See "Four Stroke Revolution" in the chapter "The Amazing Twentieth Century" for more information on the intake ports of the manifold.

[5] http://www1.eere.energy.gov/hydrogenandfuelcells/tech_validation/pdfs/fcm03r0.pdf

[6] http://en.wikipedia.org/wiki/Spark_plug
http://www1.eere.energy.gov/hydrogenandfuelcells/tech%5Fvalidation

[7] http://fueleconomytips.com/2005/12/13/spark-plug-comparison-test/

[8] $2H_2 + O_2 = 2H_2O$

[9] $N_2 + NOx$

[10] http://www.ford.com/en/innovation/engineFuelTechnology/hydrogenInternalCombustion.htm The Model U was meant to evoke memories of the Model T, an automotive milestone.

[11] http://www.autointell~News~2003/August-2003/August-2

[12] Gram = 0.03527 ounces. 128 ounces = 1 gallon (U.S.).

[13] The difference in efficiency depends on the efficiency of the particular internal combustion engine, which can vary between machines and during operation.

END NOTES

[14] http://www.autointell-news.com/News-2003/August-2003/August-2003-2/August-13-03-p1.htm

17. THE PEM

[1] http://www.fctec.com/fctec_history.asp

As with a number of other discoveries, there are historians who may dispute Grove's claim. For instance, in 1838 German scientist Christian Friedrich Schönbein is said to have discovered the principle of the fuel cell and published an account in the January 1839 edition of the "Philosophical Magazine." See http://en.wikipedia.org/wiki/Fuel_cell

[2] Electrolysis: $2 H_2O \rightarrow 2 H_2 + O_2$.

[3] Reverse Electrolysis: $2H_2 + O_2 \rightarrow 2H_2O$.

[4] http://www.americanhistory.si.edu/fuelcells/origins/orig1.htm

[5] Multiple references:

http://www.powergeneration.siemens.com/en/fuelcells/history/index.cfm
http://www.fuelcellstore.com/products/heliocentris/INTRO.pdf
http://www.scied.science.doe.gov/nmsb/hydrogen/Guide%20to%20Fuel%20Cells.pdf
http://media.wiley.com/product_data/excerpt/7X/04708485/047084857X.pdf

[6] Author illustration – S. L. Klein.
Also see http://www.fctec.com/fctec_types_pem.asp

[7] 6 amps x 0.7 volts x 24 = 100.8 watts.

[8] Thomas Grubb may have gotten the idea to use a proton (ion) exchange membrane from water softeners. Another possibility was Fermi's use of cellophane in his atomic experiments.

[9] Daimler-Benz was Daimler Chrysler until the breakup in 2007.

[10] Author Illustration – S. L. Klein

18. HY-WIRE ACT

[1] http://www.cardesignnews.com/autoshows/2002/parts/preview/gm-hywire/

[2] http://www.evworld.com/article.cfm?storyid=657

19. STATIONARY POWER FUEL CELLS

[1] http://www.iee.org/Policies/Areas/EnvEnergy/FUELCELLS.pdf
http://www.eea-inc.com/dgchp_reports/CEC-Market_Assessment_CHP_CA.pdf

[2] http://www.iee.org/Policies/Areas/EnvEnergy/FUELCELLS.pdf
http://www.eea-inc.com/dgchp_reports/CEC-Market_Assessment_CHP_CA.pdf

[3] http://www.iee.org/Policies/Areas/EnvEnergy/FUELCELLS.pdf
http://www.eea-inc.com/dgchp_reports/CEC-Market_Assessment_CHP_CA.pdf

[4] The amount of heat varies enormously depending on the electrolyte selected.

[5] Author Illustration – S. L. Klein

[6] There is another fuel cell invented at JPL that uses methanol instead of hydrogen. It doesn't require a reformer.

[7] http://www.nfcrc.uci.edu/2/ACTIVITIES/RESEARCH_STUDIES/Hybrid_Fuel_Cell_Systems/Analyses_of_Hybrid_FC_Gas_Turbine_Systems/Index.aspx
Also see http://www.energy.ca.gov/reports/2002-01-11_600-01-009.PDF

[8] http://www.netl.doe.gov/publications/press/2006/ti_fuelcellhybrid_startup.html

20. HYDROGEN PRODUCTION, DISTRIBUTION, AND STORAGE

[1] Electrolysis is the oldest method.

END NOTES

[2] The amount of pollution involved in manufacturing PECs depends on the electrolyte that is used.

[3] Toshiba Press Release, May 2, 2000
http://www.toshiba.co.jp/about/press/2000_05/pr0201.htm

[4] One discovery that helped advance PEC technology was announced in 2004. Australian scientists found that relatively low cost titanium dioxide (TiO_2) electrodes can enhance the effectiveness of sunlight in splitting water or another substance with the right composition into hydrogen and oxygen.

http://www.energybulletin.net/1807.html for some details.

[5] http://pubs.acs.org/cgi-bin/sample.cgi/jacsat/2006/128/i49/pdf/ja0643801.pdf Also see the 13 December issue of the Journal of the American Chemical Society.

[6] http://pubs.acs.org/cgi-bin/jcen?esthag/asap/html/es050244p.html

[7] "Electrochemically Assisted Microbial Production of Hydrogen from Acetate," by Hong Liu, Stephen Gro, and Bruce E Logan, Environmental Science Technology, 2005, 39, 4317-4320.

[8] Some would disagree that electrolysis requires 60% more energy input than it produces. My experience suggests electrolysis requires less than 40% more energy as input than it produces.

[9] http://pubs.acs.org/cgi-bin/jcen?esthag/asap/html/es050244p.html See "Electrochemically assisted Microbial Production of Hydrogen from Acetate," Envion Sci Technol 2005, 39, 41317-4320.

[10] http://pubs.acs.org/cgi-bin/jcen?esthag/asap/html/es050244p.html

[11] The discovery could add to our understanding of Earth's primitive atmosphere.

[12] Today approximately 69% of electricity is generated using coal, natural gas, and oil; 18 to 21% (depending on what government source is consulted) using nuclear energy.

[13] According to a Chinese study, if production efficiency of green algae is raised to 10%, hydrogen production can reach 18 grams per square meter per day.

http://www.newenergy.org.cn/english/solar/science/conversion/hydrogen.htm

[14] The DOE spent approximately $2 to $3M on photolytic research in 2004 for photobiological and photoelectrochemical research. The projected budget for the next year for photolytic research was $8.1 M.

[15] http://www.wired.com/science/discoveries/news/2006/02/70273

[16] The gas shift reaction is a reaction in which water and carbon monoxide react to form carbon dioxide and hydrogen.
$CO + H_2O => H_2 + CO_2$.

[17] $CH_4 + H_2O => CO + 3H_2$

[18] $F = 1.8C+32$

[19] Hydro-Gasification: $H_2O + C + 2H_2 \rightarrow CH_4 + H_2O$.

[20] Carbonation: $CH_4 + 2H_2O\tilde{\ } = CO_2 + 4H_2$; $CO_2 + CaO \rightarrow CaCO_3$

[21] Calcination: $CaCO_3 \rightarrow CO_2 + CaO$

[22] Illustration – S. L. Klein

[23] See "Developing an Alternative Infrastructure" for more information on electrochemical cells.

[24] Author Illustration – S. L. Klein

[25] Various methods of extracting hydrogen are covered in the chapter on "Building an Alternative Infrastructure."

[26] $2H_2 + CO_2 = CH_4 + O_2$

[27] http://fsi.ucf.edu/special/SolarSystem/mars/TC/TC010.pdf

[28] http://spot.colorado.edu/~meyertr/rwgs/rwgs.html
Chemists have been aware of the reverse water gas shift (RWGS) reaction since the mid 1800's.

[29] $2H_2 + CO + O_2$

[30] For example, $2H_2 + CO + O_2 = CH_3OH + O_2$

[31] See Steven Gloor, *A proposal to create a 21st Century economy based on renewable, man-made methane.*

http://www.stevegloor.typepad.com/sgloor/files/methane_economy.pdf
http://www.evworld.com/view.cfm?section=article&storyid=781

[32] See Zubrin, Robert et al. *Report on the Construction and Operation of a Mars Methanol in situ Propellant Production Unit*

END NOTES

http://www.pioneerastro.com/Projects/Mmispp/MMISP_Phae_1_Final_Report.doc

[33] *Hydrogen Posture Plan An Integrated Research, Development, and Demonstration Plan*, U.S. Department Of Energy, February 2004, pg. ii

[34] See *National Hydrogen Energy Roadmap*, November 2002, U.S. Department of Energy pg. 11.

[35] Hydrogenization of fats produces trans-fatty acids which are now considered a health threat.

[36] *A National Vision of America's Transition to a Hydrogen Economy - To 2030 and Beyond*, November 2001, U.S. Department of Energy, pg 4.

[37] You may recall that an oil company Chevron got involved after GM decided to sell its interest in Ovonics' NiMH battery. Chevron eventually took over Texaco and inherited Ovonics' battery

[38] http://www.globeinvestor.com/servier

[39] *Recent Advances in Solid Hydrogen Storage Systems*, Chao, B.S. et al. Texaco Ovonics Hydrogen Systems. For more information, see http://www.ovonic.com for more details.

[40] See *National Hydrogen Energy Roadmap*, November 2002, U.S. Department of Energy pg. 18.

[41] For more information on federal laws and regulations governing hydrogen, see Title VIII- Hydrogen

http://energy.senate.gov/legislation/energybill2003/hydrogen3.pdf

21. TECHNOLOGICAL READINESS ASSESSMENT

None.

22. THE LACK OF VISIBLE PROGRESS

[1] *A National Vision of America's Transition to a Hydrogen Economy - To 2030 and Beyond*, November 2001, U.S. Department of Energy.

[2] *National Hydrogen Energy Roadmap*, November 2002, U.S. Department of Energy.

[3] *Hydrogen Posture Plan An Integrated Research, Development, and Demonstration Plan*, February 2004, U.S. Department of Energy. http://www.eere.energy.gov/hydrogenandfuelcells/posture_plan04.html

[4] *A National Vision of America's Transition to a Hydrogen Economy To 2030 and Beyond*, page i.

[5] *A National Vision of America's Transition to a Hydrogen Economy - To 2030 and Beyond*, pages iii-iv.

[6] *A National Vision of America's Transition to a Hydrogen Economy - To 2030 and Beyond*, page 14.

[7] *Hydrogen Posture Plan An Integrated Research, Development, and Demonstration Plan*, , pg. 8 for a summary of needs and challenges.

[8] *Hydrogen Posture Plan An Integrated Research, Development, and Demonstration Plan*, pg. 2

[9] *Hydrogen Posture Plan An Integrated Research, Development, and Demonstration Plan*, pg. 21.

23. WHAT HOLDS US BACK

[1] "More Than 15,000 Scientists Protest Kyoto Accord; Speak Out Against Global Warming Myth"

http://www.sepp.org/pressrel/petition.html

[2] *Global Warming Hoax – The Anti-Capitalist Scheme* by Gerry Gannon & G.E. Krekelberg, http://www.thecapitalist.net/globwarm.html

[3] Base, Joseph et al., *Eco-Sanity: A Common Sense Guide to Environmentalism*, June 1996.

[4] Caruba, Alan, *The Great Hydrogen Myth*. February 11, 2003. See http://www.cnsnews.com/ViewPrint.asp?Page=\Commentary\archive\20 0302\COM20030211a.html

[5] Kunstler, James, "The Long Emergency," Rolling Stone, March 2005. See http://www.countercurrents.org/po-Kunstler280305.htm. He also has a book on the same subject.

[6] Kunstler, James, "The Long Emergency."

END NOTES 343

[7] 1 unit output /1 unit input = no gain or loss; 1 unit output /2 units input = a loss; 1 unit output / half a unit input = an energy gain;

[8] See the study *PV Payback* by Karl E Knapp & Thereas L Jester, http://www.homepower.com/files/pvpayback.pdf

[9] It's free until the government finds a way to tax it.

[10] http://www.nrel.gov/ncpv/hotline/09_00_siemens; and http://www.homepower.com/files/pvpayback.pdf

[11] See Union of Concerned Scientists http://www.ucsusa.org/CoalvsWind/brief.coal.html

[12] http://obsidianwings.blogs.com/obsidian_wings/2005/03/sunny_news_for_.html

[13] http://www.ucsusa.org/clean_energy/coalvswind/brief_coal.htm for more details

[14] http://obsidianwings.blogs.com/obsidian_wings/2005/03/sunny_news_for_.html

[15] *The Main Problem with Alternatives to Oil*, 4 June 2005, http://www.eclipsenow.org/Facts/alternateenergy.html

[16] http://www.seas.ucla.edu/hsseas/releases/blimp.htm
http://engineer.ea.ucla.edu/releases/blimp.htm for more details.

[17] http://web.archive.org/web/20041011071632/http://engineer.ea.ucla.edu/releases/blimp.htm

[18] Behar, Michael, "Warning: The Hydrogen Economy May Be More Distant Than It Appears"

http://www.freerepublic.com/focus/f-news/1301843/posts

[19] Behar, Michael, "Warning: The Hydrogen Economy May Be More Distant Than It Appears"

[20] Behar, Michael, "Warning: The Hydrogen Economy May Be More Distant Than It Appears"

[21] http://www.eere.energy.gov/de/power_crunch.html

[22] Energy Information Administration Monthly Energy Review, Section 9. Energy Prices, January 2006. See http://www.eia.doe.gov/emeu/mer/pdf/mer.pdf

[23] http://findarticles.com/p/articles/mi_m2744/is_3_2007/ai_n19001202

[24] http://www.eia.doe.gov/neic/infosheets/petprices.html

[25] http://www.exxonmobil.co.uk/UKEnglish/Newsroom/UK_NR_Speech_EO_150904.asp

24. CALIFORNIA BELLWETHER

[1] http://en.wikipedia.org/wiki/California_electricity_crisis for more details.

[2] http://www.iea.org/textbase/speech/2003/phbilling.pdf

[3] For a chronology of California's energy crisis
http://en.wikipedia.org/wiki/California_electricity_crisis
http://www.pbs.org/wgbh/pages/frontline/shows/blackout/california/
http://www.ferc.gov/industries/electric/indus-act/wec.asp

[4] http://feinstein.senate.gov/03Speeches/engcr1081b.htm

[5] http://www.cbsnews.com/stories/2004/06/08/eveningnews/

[6] http://www.marketwatch.com/News/Story/Story.aspx?guid=%7B4061B1B0-7DC7-4A4F-AE4A-3C119D69A93A%7D&siteid=mktw and http://www.marketwatch.com/search/?value=Jason+Leopold+California+Blackouts

[7] http://www.marketwatch.com/search/?value=Jason+Leopold+California+Blackouts page.2 and http://www.marketwatch.com/News/Story/Story.aspx?guid=%7B4061B1B0%2D7DC7%2D4A4F%2DAE4A%2D3C119D69A93A%7D&siteid=mktw

[8] http://www.hempfarm.org/Papers/Enron_vs_Calif.html

[9] http://www.msnbc.msn.com/id/5798109 ,AB 2076 Report 2003P600-03-005F).

[10] A group called *Energy Independence Now* tracks legislative bills in California related to hydrogen.

END NOTES   345

See http://www.energyindependencenow.org/legistlative.html

[11] The U.S. oil reserve can hold around 700 million barrels. Americans consume twenty-one million barrels a day. Around 6 out of every 10 barrels of oil are imported.

25. THE FUTURE IN OUR HANDS

[1] http://www.epinet.org/content.cfm/webfeatures_econindicators_tradepict20050210

[2] http://www.epinet.org/content.cfm/webfeatures_econindicators_tradepict20050210

[3] Author illustration – S. L. Klein
See http://www.eia.doe.gov/cneaf/electricity/epa/epa_sum.html and http://www.eia.doe.gov/kids/infocardnew.html

[4] Some sources show significant increases in alternative energy in the stationary power sector in recent years (since 2004). Others do not.
   http://www.eia.doe.gov/kids/.../CrunchTheNumbersIntermediateDec2002.pdf
   http://www.blm.gov/nhp/pubs/brochures/EnergyBro.htm
   http://www.rff.org/rff/News/Coverage/2004/June/.../getfile.cfm&PageID=14448
   http://www.csbsju.edu/environmentalstudies/curriculum/energy/Introduction.htm

[5] A few states are already using ethanol.

[6] An ethanol-gasoline blend is sometimes called gasohol.

[7] A measure relies on a single parameter such as "the distance a car can go on a "tank" of fuel. A metric relates two or more parameters. An example of a metric is average miles per gallon verses number of barrels of oil imported.

[8] "Energy Efficiency is the Best Security" by Harvey Michaels @ http://www.nexusenergy.com/presentation5.aspx

[9] Energy information agency statistics and graph at www.eia.doe.gov/emeu/mer/pdf/pages/sec9.pdf

[10] Remember the estimate that a 5% lowering of demand would have resulted in a 50% price reduction during the peak hours of California's electricity crisis of 2000 and 2001. The same is likely to be true for other states. California has gone far in addressing ways to forestall blackouts. Other states can find much of value in California's actions.

26. THE NEAR-TERM CAMPAIGN

[1] Author illustration – S. L. Klein

[2] Among the references are Energy Information Administration (EIA), the Bureau of Transportation Statistics (BTS), Exxon Mobil and British Petroleum (BP), Gibson Consulting Oil Statistics, Marcon International Inc., Rand, peakoil.com, and various news organizations including BBC and CNN. In particular I reference:

http://www.eia.doe.gov/neic/quickfacts/quickoil.html
http://www.eia.doe.gov/neic/quickfacts/quickelectric.html
http://www/eoa/dpee,eia/aer/txt/prb0802a.html
http://en.wikipedia.org/wiki/Passenger_vehicles_in_the_United_States
http://www.bts.gov/publications/national_transportation_statistics/2006/html/table_01_11.html
http://en.wikipedia.org/wiki/World_population
http://www.fueleconomy.gov/feg/calculator_selectengine.jsp?year
http://www.epa.gov/otaq/cerl/mpg/fetrends/420r07008.pdf

[3] As a snapshot, British Petroleum indicated U.S. production in 2004 was about 7.4 million barrels per day. See BP Statistical Review of World Energy June 2006. The U.S. Energy Information Administration / Annual Energy Review 2005 reports U.S. production was 5.4 million barrels per day in 2004. Gibson Consulting Oil Statistics put U.S. crude oil production capacity at about 5 million barrels per day (mb/d) in 2004. The Bureau of Transportation Statistics reported in 1993 that the net import of petroleum was 7.63 mb/d. The Energy Information Administration in its Annual Energy Review 2005 Table 5.1 put net imports of petroleum in 1993 at 7.618 mb/d.

[4] http://www.eia.doe.gov/kids/energyfacts/sources/non-renewable/oil.html#Howused A graph shows a 42 gallon barrel produces 20 gallons of gasoline and 7 gallons of diesel fuel or around 64% of the

END NOTES

total. In an L.A. Times article, May 10, 1985 Patrick Boyle mentions that 18 gallons of gasoline was derived per 42 gallon barrel of oil.

For more details, see
www.ftc.gov/reports/gasprices05/050705gaspricesrpt.pdf
http://www.bts.gov/publications/national_transportation_statistics/html/appendix_c_chapter_04.html
http://www.atg.wa.gov/Antitrust/GasPrices/GasolineReport.aspx

[5] See Petroleum Overview, Selected Years 1949-2005 and Electricity Overview, from the Energy Information Administration Annual Energy Review 2005 and Basic Petroleum Statistics October 2006 of Energy Information Administration.

Several other sources were also used including:

http://www.npg.org/fact/sworld_pop_year.htm
http://www.census.gov/prod/20054pubs/03statablpop.pdf

Energy Information Administration Table D1 U.S. Proven Resources of Crude Oil 1975-2004 Energy Information Administration Table D1 U.S. Proven Resources of Crude Oil 1975-2004.

See Table 8.1 Electricity Overview, Selected Years 1949-2005 from the Energy Information Administration Annual Energy Review 2005
http://www.eia.doe.gov/emeu/aer/txt/ptb0809.html

[6] 8.2 billion MWh per year x 0.45 = 3.7 billion MWh per year.

[7] See "Using Coal Gasification and Fracturing to Obtain Oil" in Chapter 13 for the Fischer-Tropsch and low-Temperature Carbonization (LTC) conversion processes.

[8] http://news.mongabay.com/2006/0816-wsj.html

http://www.moneyweek.com/file/13377/could-coal-replace-oil.html

[9] http://news.mongabay.com/2006/0816-wsj.html

[10] http://www.ktvu.com/globalwarming/10086689/detail.html
http://www.eia.doe.gov/oiaf/servicerpt/powerplants/chapter2.html
http://www.findarticles.com/p/articles/mi_qn4196/is_20060630/ai_n16517883

[11] http://www.moneyweek.com/file/13377/could-coal-replace-oil.html
http://news.bbc.co.uk/2/hi/programmes/newsnight/4330469.stm
http://www.csmonitor.com/2007/0322/p01s04-wogi.htm

[12] http://www.fossil.energy.gov/programs/powersystems.gassification/
http://www.moneyweek.com/file/13377/could-coal-replace-oil.html

[13] http://www.fossil.energy.gov/programs/powersystems.gassification/
http://www.coal2oil.com/
http://www.moneyweek.com/file/13377/could-coal-replace-oil.html

[14] http://www.greencarcongress.com/2006/04/rumors_rampant_.html
http://www.news.com/8301-10784_3-9755379-7.html

[15] This scenario is one of many that could be devised to meet intervention goals.

[16] http://www.time.com/time/printout/0,8816,914582,00.html
http://www.history.com/encyclopedia.do?articleId=201868

[17] http://www.bookrags.com/Energy_intensity

[18] http://en.wikipedia.org/wiki/Corporate_Average_Fuel_Economy
http://uspolitics.about.com/od/energy/l/bl_cafe_table.htm
http://www.ce.utexas.edu/prof/kockelman/public_html/LDTpaper_to_Transp.pdf
http://www.aceee.org/briefs/menu2.htm
http://www.nhtsa.dot.gov/cars/rules/cafe/overview.htm
http://www.encyclopedia.com/doc/1G1-102662135.html
http://www.fueleconomy.gov/feg/ratings2008.shtml

[19] Calculation: 17.6 mb/d x 0.70 = 12.32 mb/d is the amount consumed by light duty vehicles with a 21 mpg average. The amount consumed if light duty vehicles get 51 mpg is 12.32 mb/d x 0.41= 5.05 mb/d, which is a savings of 12.32 mb/d - 5.05 mb/d = 7.3 mb/d .

[20] http://www.energybulletin.net/21064.html
http://www.eia.doe.gov/emeu/cabs/Barcil.pdf.pdf

[21] http://en.wikipedia.org/wiki/Ehtanel_fuel_in_Brazil

END NOTES 349

[22] http://www.wastemanagement.com/wm/environment/documents/Environmental_ Review.pdf

[23] http://www.recycleamerica.com
http://www1.eere.energy.gov/biomass
http://www.eia.doe.gov/kids/energyfacts/saving/recycling/solidwaste/recycling.html "Americans recycled just six percent in 1960...but we recycle about 31% today."

[24] http://www.recycleamerica.com/

[25] If 1/3 of drivers reduced their driving by one day a week by carpooling or telecommuting, then the percentage of gasoline saved would be roughly 1/7 x 1/3 x 100% = 5%. Because light duty vehicles would be consuming 5.05 mb/d (see footnote 18), a 5% decrease in consumption would be 5.05 x 0.05 ≈ 0.3 mb/d.

[26] http://www.financialsense.com/editorials/cooke/2005/0412.html

[27] http://www.pdfpeakoil.blogspot.com/2005/01/chinas-risky-scramble-for-oil.html
http://www1.eere.energy.gov/industry/petroleum_refining/pdfs/petroleumroadmap "The average American consumes 20 pounds of petroleum a day." Note that a barrel of oil weighs 320 pounds.

[28] 0.1mb/d = 100,000b/d. 100,000/300 = 333.33 b/d; 8y*0.25 +1y = 333.33; y =111. 8y= 888. 888(0.25)+111 = 333

[29] http://www.energystar.gov

[30] See Energy Star @ http://www.energystar.gov/
http://www.energystar.gov/index.cfm?c=appliances.pr_appliances
http://www.energystar.gov/index.cfm?c=heat_cool.pr_hvac
http://www.eere.energy.gov/consumer/your_home/space_heating_cooling/index.cfm/mytopic=12370

[31] Based on Energy Star Pie Chart.
http://www.energystar.gov/index.cfm?c=products.pr_pie

[32] http://www.bosgraaf.com/homes/es1000.html "An Energy Star home is 40 to 50 percent more efficient than new homes built to the 1993

National Model Energy Code. Building to Energy Star standards is not complex or highly expensive."

[33] http://www.energy.gov/energyefficiency/5790.htm

[34] http://www.tec.appstate.edu/construction/passive_solar/
http://www.builditsolar.com/Projects/Cooling/passive_cooling.htm

[35] http://www.dreamgreenhomes.com/aboutus/profiles.htm
http://www.electronichouse.com/article/energy_monitoring_in_the_home/

[36] Assume homes 20 years from now are super efficient - 55% more efficient than the average home today. We have gotten consumption down to 8.2-3.7 = 4.5 billion megawatt-hours per year. To get it back to 4.1 we need to increase generation of electricity using alternative energy by 0.4 billion megawatt-hours per year or (4.5-0.4)/4.5= 91% of the total needed consumption.

http://www.eere.energy.gov/news/archives/2001/nov07_01.html
http://www.eia.doe.gov/aer/txt/ptb1001.html
http://www.eia.doe.gov/oiaf/ieo/electricity.html

[37] The more our population grows, the more it will jeopardize the possibility of succeeding. The faster we can accomplish these goals the easier it will be to achieve them given continued population growth.

[38] http://www.calcars.org/vehicles.html

[39] http://www.greencarcongress.com/2006/12/doe_study_offpe.html

[40] See footnote 18.

[41] See footnote 24.

[42] See EIA Annual Energy Review Table 5.1.2a-b or
http://www.eia.doe.gov/pub/oil_gas/petroleum/analysis_publications/oil_market_basics/dem_image_us_cons_sector.htm

[43] http://www.fossil.energy.gov/programs/powersystems/gasification/howgasificationworks.html

END NOTES

[44] http://www.fossil.energy.gov/programs/powersystems/gasification/indeix.html

[45] http://en.wikipedia.org/wiki/World_population

27. THE ALTERNATIVE ENERGY ZONE

[1] See Chapter One.

[2] It would be preferable not to provocatively advertise these efforts at a national and hence international level. But without national attention focused on the problems we have with energy, we jeopardize success.

[3] Author diagram – S.L. Klein

[4] http://www.eere.energy.com/de/minigrids.html

[5] Example is drawn from actual California Edison electric bills. Note that seasonally and for other reason Edison varies it charges so that in November 382 kilowatt hours cost less than 366 kilowatt hours in February. I naturally checked to make sure this was correct. It was.

[6] http://www.eere.energy.gov/consumer/your_home/space_heating_cooling/index.cfm/mytopic=12370

[7] http://www.amazon.com/gp/product/B0001EY6P0/102-2109888-4397707?v=glance&n=228013

[8] http://www.mge.com/images/PDF/Brochures/Community/ApplianceEnergyCost.pdf

[9] Check Energy Star website http://www.energystar.gov for more information.

[10] http://macdailynews.com/index.php/weblog/comments/5006/

[11] http://www.southwestpv.com/Catalog/Home%20Power/Net-metering.htm

[12] This list is based on currently existing systems.

[13] http://www.nrel.gov/docs/fy04osti/35297.pdf
http://www.southwestpv.com/Catalog/Home%20Power/Net-metering.htm#System1
http://projectsol.aps.com/solar/data_insolation.asp

[14] http://www.southwestpv.com/Catalog/Home%20Power/Net-metering.htm#System1

[15] The power generation needed during 5.5 hours of sun exposure to produce and store electricity for the whole day is 15 kWh / 5.5 hours = 2.7 kW. A solar panel produces 70 mW/in^2 = 0.000070 kW/in^2. Thus, the 2.7 kW can be produced by a solar panel with an area of 2.7 kW / (0.000070 kW/in^2) ≈ 39,000 in^2 ≈ 270 square feet.

[16] http://rredc.nrel.gov/wind/pubs/atlas/maps/chap2/2-06m.html
http://www.nrel.gov/wind/wind_map.html

[17] http://www.utilityfree.com/wind/windpower.html

[18] Wind turbines that can generate electricity at lower speeds exist but tend at present to be very expensive.

[19] Wind efficiency is determined by dividing the power curve by the area of the rotor to get the power output per square meter (m^2) of rotor area. The power curve of a wind turbine is a graph that indicates the electrical power output for a specific turbine at different wind speeds (W). For additional information see

http://www.windpower.org/en/tour/wres/cp.htm

http://www.healthgoods.com/Education/Energy_Information/Renewable_Energy/small_wind_systems.htm

[20] NREL indicate that sites that are in class 2 (Marginal – approximately 12.5 mph) or higher can be used for small wind generators. Areas that have wind conditions in Class 3 (Fair – approximately 14.3 mph) or higher can be considered as candidates for wind farms.

[21] http://www.etaengineering.com/windpower/h80.shtml

[22] http://home.golden.net/~red/wind_turbines.htm

[23] http://www.southwestpv.com/Catalog/Wind%20Power/H80.HTM

[24] 930/250 = 3.7; 930/2 = 465; 465/250 = 1.9

http://www.southwestpv.com/Catalog/Wind%20Power/H80.HTM
http://windside.com

END NOTES

[25] Author Illustration – S. L. Klein

[26] http://dmoz.org/Science/Technology/Energy/Storage/

[27] http://home.earthlink.net/~fradella/homepage.htm

[28] http://www.greencarcongress.com/2005/04/solar_electric_.html

The University of Queensland group had approximately 2½ square meters of solar panels (27 square feet) on their car. An output of 375 W for these solar panels corresponds to 96 mW/in^2, which is more efficient than the 70 mW/in^2 of typical home solar cells. (Calculation: 375 W / 27 ft^2 x (ft/12 in)^2 x (1000 mW / W) = 96 mW/in^2)

[29] Recharging with a solar blanket at 375 W for 5.5 hours would result in 2 kWh of stored electrical energy. If the RAV4-EV uses 27.4 kWh of electrical energy to go 100 miles, then 2kWh from a solar blanket recharging would allow it to go only 7 miles.
(Calculation: 100 miles x 2kWh / 27.4 kWh = 7 miles)

[30] www.eere.energy.gov/afdc/pdfs/hyd_economy_bossel_eliasson.pdf
The estimate for how much electricity may be required to produce hydrogen also seems excessive. Time and real world experiments will tell.

[31] For an example, see Ovonics metal-hydride hydrogen storage system (OvMH) already discussed previously.

[32] If a hydrogen-electric hybrid car can go roughly 45 miles on one kilogram of hydrogen, a hydrogen fuel cartridge system that contains at least 6 kilograms would work, allowing for some additional storage weight over that of a conventional car. If the storage system adds significant additional weight to the car, then this estimate would be incorrect. Calculation: 45 miles per kilogram * 6 kilograms = 270 miles

[33] http://www.ovonic.com/PDFs/ecd_hydrogen_economy_sro_mrs_nov03/sro_mrs_hydrogen_nov2003_042704.pdf

[34] http://www.cem.msu.edu/~cem181h/projects/97/fuelcell/chem1.htm
This claim strains credulity.

[35] Assuming an absorptive material of sufficient abundance is used.

[36] See Chapter 6 The Environment, the Mad Hatter, and His Dirty Dishes.

[37] http://www.newrules.org/agri/ethanol.html

[38] http://www.energylan.sandia.gov/sunlab/research.htm

[39] http://www.osti.gov/bridge/product.biblio.jsp?osti_id=36778
Validation of the FLAGSOl Parabolic Trough Solar Power Plant Performance Model

[40] http://www.energy.ca.gov/wind/overview.html for more details

[41] See Wikipedia on line. According to Wikipedia, San Francisco's Population in 2004 was around 744,230

[42] http://www.energy.ca.gov/wind/overview.html
http://rredc.nrel.gov/wind/pubs/atlas/chp3.html

[43] http://www.nwcouncil.org/Library/1998/98-19.htm
http://www.awea.org/faq/wwt_basics.html

28. THE BOTTOM LINE

[1] Plug-ins are available from some start up companies at a cost of about $12,000 over the sticker price of a Prius. If Toyota provided plug-in capability, the cost would be, according to some estimates, $3,000.00 above the sticker price, assuming the plug-in Prius is at least as popular as the Prius.

[2] The estimate is for a family of four – Prices can vary a lot depending on what your requirements are, and when and where you look.
http://www.energystar.gov/
http://www.energy.gov/news/1282.htm

29. A CALL TO ACTION

None.

920014

Made in the USA